*Study Guide and
Problems Supplement*

ENGINEERING MECHANICS
Dynamics

SIXTH EDITION

R. C. Hibbeler

MACMILLAN PUBLISHING COMPANY
New York

MAXWELL MACMILLAN CANADA, INC.
Toronto

MAXWELL MACMILLAN INTERNATIONAL
New York   Oxford   Singapore   Sydney

Copyright © 1992, Macmillan Publishing Company,
a division of Macmillan, Inc.

Macmillan Publishing Company is part of the
Maxwell Communications Group of Companies.

Printed in the United States of America

All rights reserved. No part of this book may be reproduced or
transmitted in any form or by any means, electronic or mechanical,
including photocopying, recording, or any information storage and
retrieval system, without permission in writing from the Publisher.

Macmillan Publishing Company
866 Third Avenue, New York, New York 10022

Maxwell Macmillan Canada, Inc.
1200 Eglinton Avenue East
Suite 200
Don Mills, Ontario M3C 3N1

Printing: 1 2 3 4 5 6 7   Year: 2 3 4 5 6 7 8

ISBN   0-02-354454-6

Preface

Dynamics

This workbook is a supplement to the textbook *Engineering Mechanics: Dynamics.* Consequently, the problems in this book are arranged in an order which corresponds with the topics presented in the textbook. It will be noticed that the solution to most of the problems are only partially complete. Application of the key equations, which stress the important fundamentals of the problem solution, must be filled in by the student in the space provided. There is no need for calculations, however, all the answers are given in the back of the book.

For best results, it is intended that the problems be solved *just after* the theory and example problems covering the corresponding topic have been studied in the textbook. If an honest effort is made at completing and understanding the solution to these problems, it will serve to build confidence in applying the theory to the problems in the textbook. Furthermore, these problems provide an excellent review of the subject matter, which can then be used when preparing for exams.

The involvement of filling in part of the solution to these problems is based on a thought once expressed by Confucius:

> I *hear* and I *forget*,
>
> I *see* and I *remember*,
>
> I *do* and I *understand*

Consequently, the student who persistently solves problems in mechanics, both in this book and in the textbook, will ultimately gain a thorough, usable knowledge of mechanics, a lasting knowledge since it is based on "doing."

R.C. H.

Contents

Dynamics

12 Kinematics of a Particle 1
Rectilinear Kinematics 1
Graphic Solutions 6
Curvilinear Motion: Rectangular Components 11
Curvilinear Motion: Normal and Tangential Components 16
Curvilinear Motion: Cylindrical Components 21
Absolute-Dependent-Motion Analysis of Two Particles 26
Relative-Motion Analysis of Two Particles Using Translating Axes 29

13 Kinetics of a Particle: Force and Acceleration 35
Equations of Motion: Rectangular Coordinates 35
Equations of Motion: Normal and Tangential Coordinates 42
Equations of Motion: Cylindrical Coordinates 46

14 Kinetics of a Particle: Work and Energy 51
Principle of Work and Energy 51
Power and Efficiency 57
Conservation of Energy Theorem 61

15 Kinetics of a Particle: Impulse and Momentum 67
Principle of Linear Impulse and Momentum 67
Conservation of Linear Momentum for a System of Particles 73
Impact 79
Angular Momentum 84

16 Planar Kinematics of a Rigid Body 89
Rotation About a Fixed Axis 89
Absolute General Plane Motion Analysis 94
Relative-Motion Analysis: Velocity 99
Instantaneous Center of Zero Velocity 104
Relative-Motion Analysis: Acceleration 109

17 Planar Kinetics of a Rigid Body: Force and Acceleration 115
Equations of Motion: Translation 115
Equations of Motion: Rotation About a Fixed Axis 121
Equations of Motion: General Plane Motion 126

18 Planar Kinetics of a Rigid Body: Work and Energy 131
Principle of Work and Energy 131
Conservation of Energy 136

19 Planar Kinetics of a Rigid Body: Impulse and Momentum 141
Principle of Impulse and Momentum 141
Conservation of Momentum 148

Answers 153

12 Kinematics of a Particle

Rectilinear Kinematics

12-1. A car is traveling at a speed of 80 ft/s when the brakes are suddenly applied, causing a constant deceleration of 10 ft/s². Determine the time required to stop the car and the distance traveled before stopping.

Solution

$$v_0 = 80 \text{ ft/s}$$

Determine the time t

$$\overset{+}{\rightarrow} \quad v = v_0 + a_c t$$

───────────────────────────

$t = 8 \text{ s}$ \hfill Ans.

Determine the distance traveled

$$\overset{+}{\rightarrow} \quad v = v_0^2 + 2a_c(s - s_0)$$

───────────────────────────

$s = 320 \text{ ft}$ \hfill Ans.

Also, apply the following equation to determine the distance:

$$\overset{+}{\rightarrow} \quad s = s_0 + v_0 t + \frac{1}{2} a_c t^2$$

───────────────────────────

$s = 320 \text{ ft}$ \hfill Ans.

12-2. A particle is moving along a straight line through a fluid medium such that its speed is measured as $v = (2t)$ m/s, where t is in seconds. If it is released from rest at $s = 0$, determine its position when $t = 3$ s.

Solution

$$v = \frac{ds}{dt} = 2t$$

Set up the integral to determine $s = f(t)$.

$s = t^2$

At $t = 3$ s,

$s = (3)^2 = 9$ m *Ans.*

$$a = \frac{dv}{dt}$$

$a = 2$ m/s² *Ans.*

12-3. A ball is thrown vertically upward from the top of a building with an initial velocity of $v_A = 35$ ft/s. Determine (a) how high above the top of the building the ball will go before it stops at B, (b) the time t_{AB} it takes to reach its maximum height, and (c) the total time t_{AC} needed for it to reach the ground at C from the instant it is released.

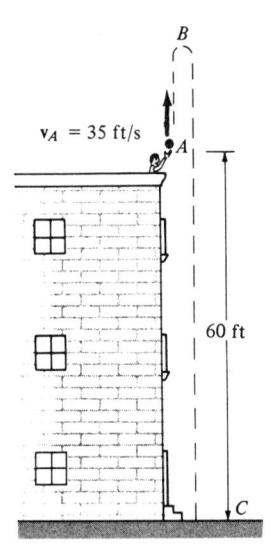

Solution

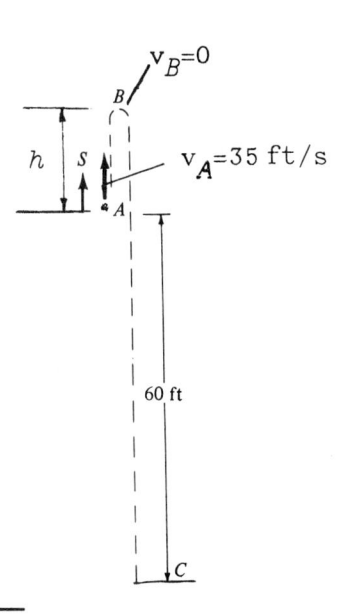

a) Determine the height h.

$$+\uparrow v_B^2 = v_A^2 + 2a_c(s_B - s_A)$$

$h = 19.0$ ft *Ans.*

b) Determine the time t_{AB}

$$+\uparrow v_B = v_A + a_c t$$

$t = 1.09$ s *Ans.*

c) Determine the time t_{AC}.

$$+\uparrow s_C = s_A + v_A t + \frac{1}{2} a_c t^2$$

$t = 3.30$ s *Ans.*

4 Study Guide and Problems

12-4. A small metal particle passes downward through a fluid medium while being subjected to the attraction of a magnetic field such that its position is observed to be $s = (15t^3 - 3t)$ mm, where t is measured in seconds. Determine (a) the particle's displacement from $t = 2$ s to $t = 4$ s, and (b) the velocity and acceleration of the particle when $t = 5$ s.

Solution

a)

$$s = 15t^3 - 3t$$

at

$$t = 2 \text{ s}, \ s = 114 \text{ mm}$$

$$t = 4 \text{ s}, \ s = 948 \text{ mm}$$

The displacement is therefore

$$\Delta s = \underline{\hspace{6em}} = 834 \text{ mm} \qquad \qquad Ans.$$

b) Determine $v = f(t)$

$$v = \frac{ds}{dt} = \underline{\hspace{6em}}$$

at

$$t = 5 \text{ s}, \ v = 1122 \text{ mm/s} \qquad \qquad Ans.$$

Determine $a = f(t)$

$$a = \frac{dv}{dt} = \underline{\hspace{6em}}$$

at

$$t = 5 \text{ s}, \ a = 450 \text{ mm/s}^2 \qquad \qquad Ans.$$

12-5. A car, initially at rest, moves along a straight road with constant acceleration such that it attains a velocity of 60 ft/s when $s = 150$ ft. Then after being subjected to *another* constant acceleration, it attains a final velocity of 100 ft/s when $s = 325$ ft. Determine the average velocity and average acceleration of the car for the entire 325-ft displacement.

Solution

Determine the first acceleration

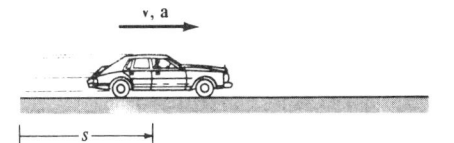

$$\stackrel{+}{\rightarrow} v_2^2 = v_1^2 + 2a_1 s$$

_____

$a_1 = 12 \text{ ft/s}^2$

Determine the second acceleration

$$\stackrel{+}{\rightarrow} v_3^2 = v_2^2 + 2a_2 s$$

_____

$a_2 = 18.29 \text{ ft/s}^2$

Determine the first time period

$$\stackrel{+}{\rightarrow} v_2 = v_1 + a_1 t_1$$

_____

$t_1 = 5 \text{ s}$

Determine the second time period

$$\stackrel{+}{\rightarrow} v_3 = v_2 + a_2 t_2$$

_____

$t_2 = 2.19 \text{ s}$

Thus

$$v_{\text{Avg}} = \frac{\Delta s}{\Delta t} = \underline{} = 45.2 \text{ ft/s} \qquad \text{Ans.}$$

$$a_{\text{Avg}} = \frac{\Delta v}{\Delta t} = \underline{} = 13.9 \text{ ft/s}^2 \qquad \text{Ans.}$$

Graphical Solutions

12-6. A car travels up a hill with the speed shown. Compute the total distance the car moves until it stops ($t = 60$ s). Plot the a-t graph.

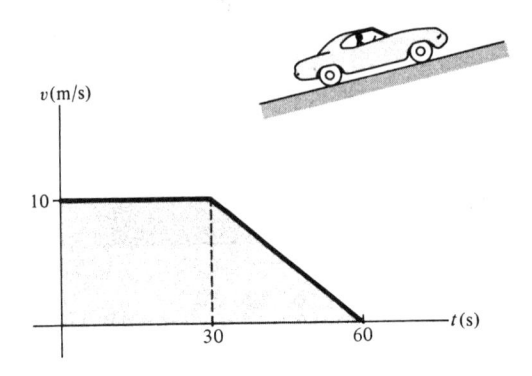

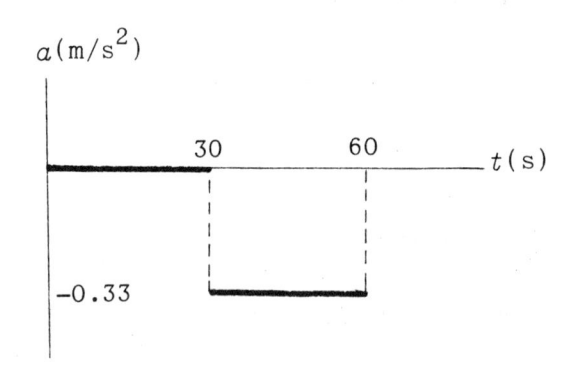

Solution

Distance traveled is area under the graph $0 \leqslant t \leqslant 60$ s.

$$A_1 + A_2 = \underline{\hspace{5cm}}$$

$s = 450$ m *Ans.*

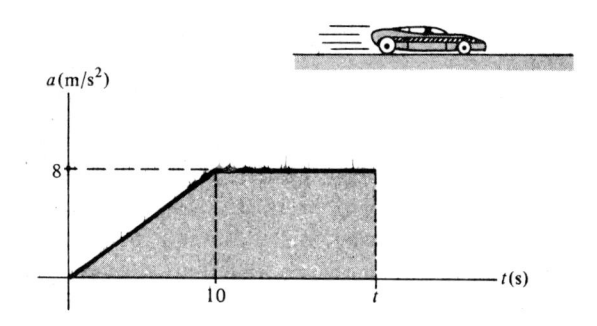

12.-7. A race car starting from rest moves along a straight track with an acceleration as shown, where $t \geqslant 10$ s, $a = 8$ m/s². Determine the time t for the car to reach a speed of 50 m/s and construct the v-t graph that describes the motion until the time t.

Solution

From the graph, for $0 \leqslant t \leqslant 10$ s

$$a = \frac{dv}{dt} = \frac{8}{10} t$$

Set up the integrals to determine $v = f(t)$

$$v = \frac{8}{20} t^2$$

At $t = 10$ s, $v = \frac{8}{20}(10)^2 = 40$ m/s

For $t > 10$ s,

$$a = \frac{dv}{dt} = 8$$

Set up the integrals to determine $v = f(t)$

$$v - 40 = 8t - 80$$

$$v = 8t - 40$$

When $v = 50$ m/s

$$t = \frac{50 + 40}{8} = 11.2 \text{ s} \quad Ans.$$

12-8. A two-stage missile is fired vertically from rest with an acceleration as shown. In 15 s the first stage A burns out and the second stage B ignites. Plot the v-t and s-t graphs which describe the motion of the second stage for $0 \leqslant t \leqslant 20$ s.

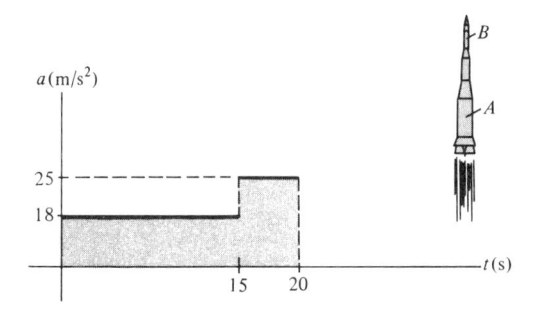

Solution

Since $v = \int a\,dt$, the constant lines of the a-t graph become sloping lines for the v-t graph. The numerical values for each point are calculated from the total area under the a-t graph to the point.

At $t = 15$ s, $v =$ _____ $= 270$ m/s

At $t = 20$ s, $v =$ _____ $= 395$ m/s

Since $s = \int v\,dt$, the sloping lines of the v-t graph become parabolic curves for the s-t graph. The numerical values for each point are calculated from the total area under the v-t graph to the point.

At $t = 15$ s, $s =$ _____

$$s = 2025 \text{ m} = 2.02 \text{ km}$$

At $t = 20$ s, $s =$ _____

$$s = 3687.5 \text{ m} = 3.69 \text{ km}$$

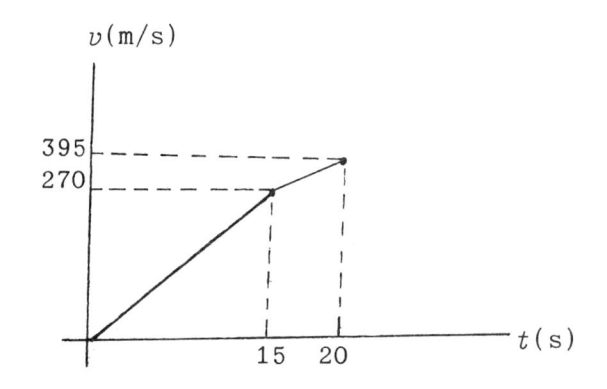

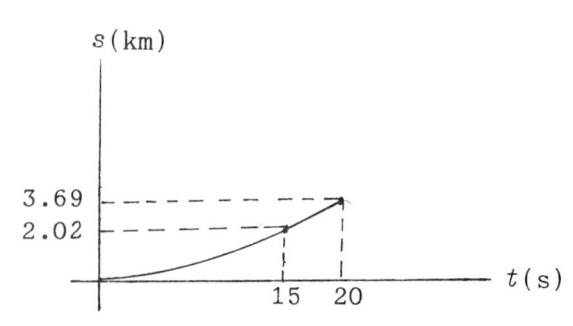

12-9. From experimental data, the motion of a jet plane while traveling along a runway is defined by the v-t graph shown. Construct the s-t and a-t graphs for the motion.

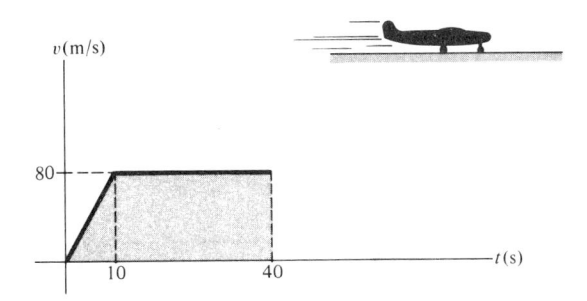

Solution

Values of a are determined from the slope of the v-t graph.

$$0 \leq t \leq 10; \quad a = \frac{\Delta v}{\Delta t} = \underline{\hspace{4in}} = 8 \text{ m/s}^2$$

$$10 \leq t \leq 40; \quad a = \frac{\Delta v}{\Delta t} = \underline{\hspace{4in}} = 0$$

The planes position at $t_1 = 10$ s and $t_2 = 40$ s is determined from the area under the v-t graph.

$$s_1 = A_1 = \underline{\hspace{4in}} = 400 \text{ m}$$

$$s_2 = A_1 + A_2 = \underline{\hspace{4in}} = 2\,800 \text{ m}$$

Set up the integrals to determine $s = f(t)$

$$0 \leq t \leq 10; \quad ds = v\,dt; \quad \underline{\hspace{4in}}$$

$$s = 4t^2$$

$$0 \leq t \leq 40; \quad ds = v\,dt; \quad \underline{\hspace{4in}}$$

$$s = 80t - 400$$

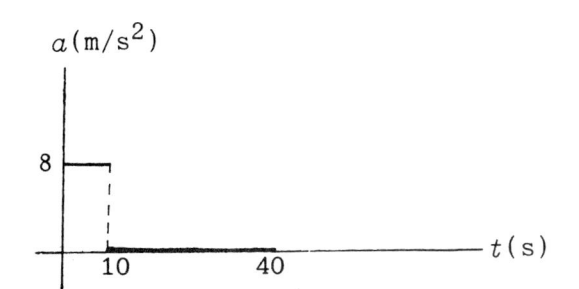

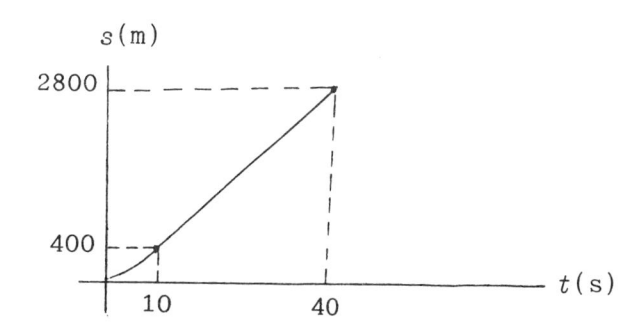

12-10. The v-s graph for a rocket sled is shown. Determine the acceleration of the sled when $s = 100$ m and $s = 175$ m.

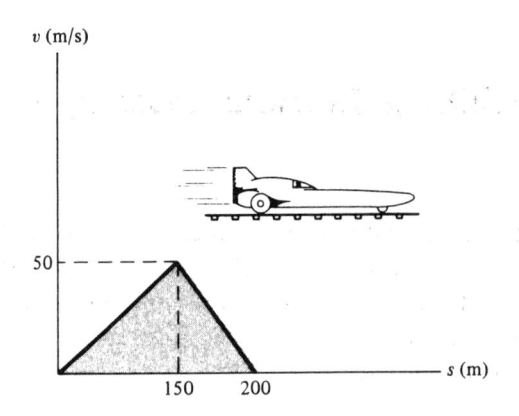

Solution

$$0 \leq s \leq 150; \quad v = \frac{1}{3}s \quad dv = \frac{1}{3}ds \quad vdv = ads$$

Set up the equation to determine $a = f(s)$.

$$a = \frac{1}{9}s$$

$$a = \frac{1}{9}(100) = 11.1 \text{ m/s}^2 \qquad \textit{Ans.}$$

$$150 \leq s \leq 200 \quad v = 200 - s \quad dv = -ds \quad vdv = ads$$

Set up the equation to determine $a = f(s)$.

$$a = s - 200$$
$$a = 175 - 200;$$

$$a = -25 \text{ m/s}^2 \qquad \textit{Ans.}$$

Curvilinear Motion: Rectangular Components

12-11. The flight path of a jet aircraft as it takes off is defined by the parametric equations $x = 1.25t^2$ and $y = 0.03t^3$, where t is the time after take-off, measured in seconds, and x and y are given in meters. If the plane starts to level off at $t = 40$ s, determine at this instant (a) the horizontal distance it is from the airport, (b) its altitude, (c) its speed, and (d) the magnitude of its acceleration.

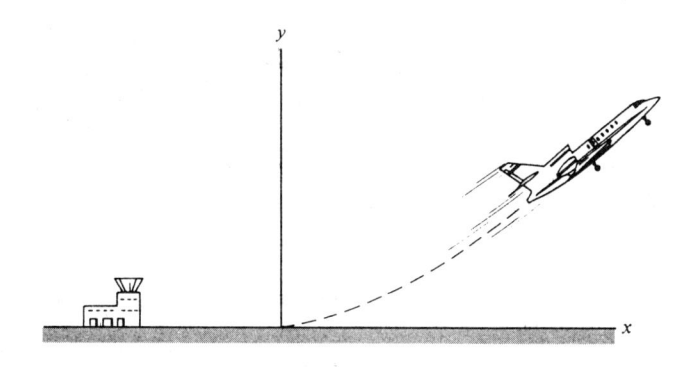

Solution

a) $x = $ _____ $= 2$ km Ans.

b) $y = $ _____ $= 1.92$ km Ans.

c) $v_x = $ _____ $= 100$ m/s

$v_y = $ _____ $= 144$ m/s

$v = \sqrt{(100)^2 + (144)^2} = 175$ m/s Ans.

d) $a_x = $ _____ $= 2.50$ m/s^2

$a_y = $ _____ $= 7.20$ m/s^2

$a = \sqrt{(2.50)^2 + (7.20)^2} = 7.62$ m/s^2 Ans.

12 Study Guide and Problems

12-12. For a short time the missile moves along the parabolic path $y = (18 - 2x^2)$ km. If motion along the ground is measured as $x = (4t - 3)$ km, where t is in seconds, determine the magnitudes of the missile's velocity and acceleration when $t = 1$ s.

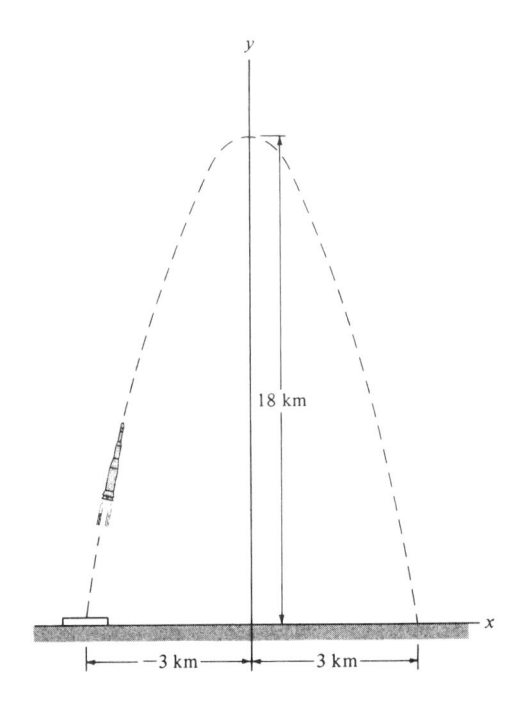

Solution

$$x = 4t - 3$$

$v_x = $ _____

$a_x = $ _____

$$y = 18 - 2x^2$$

$v_y = $ _____

$a_y = $ _____

At $t = 1$ s,

$v_x = 4$ km/s

$v_y = -16$ km/s

$v = \sqrt{(4)^2 + (-16)^2} = 16.5$ km/s   *Ans.*

$a_x = 0$

$a_y = -64$ km/s²

$a = \sqrt{(0)^2 + (-64)^2} = 64.0$ km/s²   *Ans.*

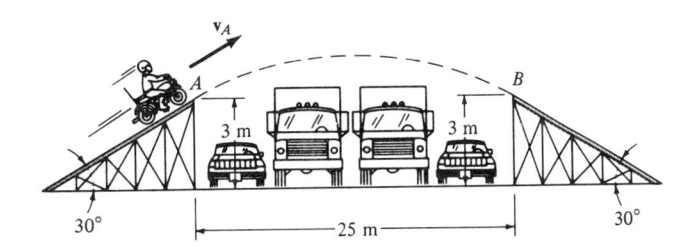

12-13. The motorcyclist attempts to jump over a series of cars and trucks and land smoothly on the other ramp, i.e., such that his velocity is tangent to the ramp at B. Determine the launch speed v_A necessary to make the jump.

Solution

Establish the x-y origin of coordinates at A. Then

$(v_A)_x = $ _____

$(v_A)_y = $ _____

Apply the following kinematic equations in the horizontal and vertical directions

$\overset{+}{\rightarrow} \; s_B = s_A + (v_A)_x t$

_____

$+\uparrow \; s_B = s_A + (v_A)_y t + \dfrac{1}{2} a_c t^2$

_____

Solving,

$v_A = 16.8$ m/s *Ans.*

12-14. The boy throws a snowball such that it strikes the wall of the building at the maximum height of its trajectory. If it takes $t = 1.5$ s to travel from A to B, determine the velocity $\mathbf{v}_A$ at which it was thrown, the angle of release θ, and the height h.

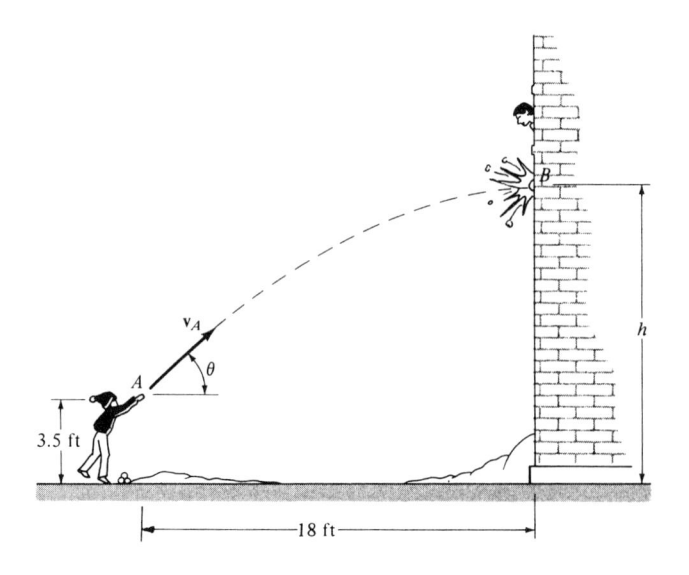

Solution

Establish the x-y origin of coordinates at A. Apply the following equations:

$$\stackrel{+}{\rightarrow} s_B = s_A + (v_A)_x t$$

$$+\uparrow (v_B)_y^2 = (v_A)_y^2 + 2a_c[(s_B)_y - (s_A)_y]$$

$$+\uparrow (v_B)_y = (v_A)_y + a_c t$$

Solving,

$$h = 39.7 \text{ ft} \qquad \qquad Ans.$$

$$\theta = 76.1° \qquad \qquad Ans.$$

$$v_A = 49.8 \text{ ft/s} \qquad \qquad Ans.$$

12-15. A ball is thrown downward on the 30° inclined plane so that when it rebounds perpendicular to the incline it has a velocity of $v_A = 40$ ft/s. Determine the distance R where it strikes the plane at B.

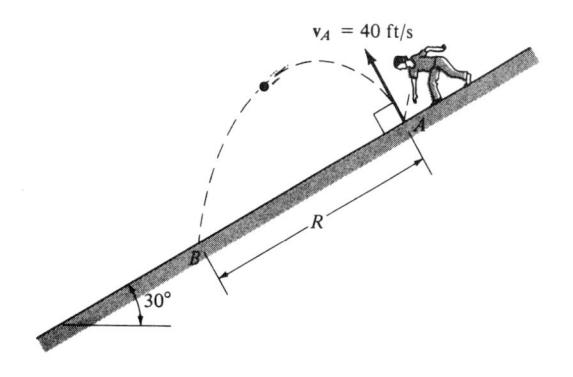

Solution

Establish the origin of the x-y coordinates at A. Then

$(v_x)_1 = $ _____ $= 20$ ft/s $\leftarrow$

$(v_y)_1 = $ _____ $= 34.64$ ft/s $\uparrow$

Write two kinematic equations in terms of R and t

$\overset{+}{\leftarrow}$ _____

$+\downarrow$ _____

Solving,

$$20t \tan 30° = -34.64t + 16.1 t^2$$

$$t = 2.87 \text{ s}$$

Thus,

$$R = \frac{20(2.87)}{\cos 30°} = 66.2 \text{ ft}$$ Ans.

Curvilinear Motion: Normal and Tangential Components

12-16. A boat is traveling along a circular path having a radius of 20 m. Determine the magnitude of the boat's acceleration if at a given instant the boat's speed is $v = 5$ m/s and the rate of increase in speed is $\dot{v} = 2$ m/s^2.

Solution

$$a_n = \underline{\hspace{2in}} = 1.25 \text{ m/s}^2$$

$$a_t = \underline{\hspace{2in}}$$

$$a = \sqrt{(1.25)^2 + (2)^2} = 2.36 \text{ m/s}^2 \qquad\qquad Ans.$$

12-17. A train travels along a horizontal circular curve that has a radius of 200 m. If the speed of the train is uniformly increased from 30 km/h to 45 km/h in 5 s, determine the magnitude of the acceleration at the instant the speed of the train is 40 km/h.

Solution

$$a_t = \frac{\Delta v}{\Delta t} = \underline{\hspace{6cm}} = 0.833 \text{ m/s}^2$$

$$a_n = \frac{v^2}{\rho} = \underline{\hspace{6cm}} = 0.617 \text{ m/s}^2$$

$$a = \sqrt{(0.833)^2 + (0.617)^2} = 1.04 \text{ m/s}^2 \qquad \textit{Ans.}$$

12-18. A sled is traveling down along a curve which can be approximated by the parabola $y = \frac{1}{4}x^2$. When point B on the runner is coincident with point A on the curve ($x_A = 2$ m, $y_A = 1$ m), the speed of B is measured as $v_B = 8$ m/s and the increase in speed is $dv_B/dt = 4$ m/s^2. Determine the magnitude of the acceleration of point B at this instant.

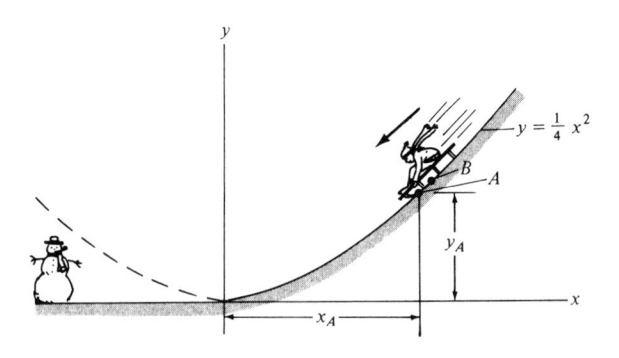

Solution

$$y = \frac{1}{4}x^2$$

$$\left.\frac{dy}{dx}\right|_{x=2} = \frac{1}{2}x = 1$$

$$\frac{d^2y}{dx^2} = \frac{1}{2}$$

$$\rho = \left| \frac{\left[1 + \left(\frac{dy}{dx}\right)^2\right]^{3/2}}{\frac{d^2y}{dx^2}} \right| = \left| \frac{[1 + (1)^2]^{3/2}}{\left(\frac{1}{2}\right)} \right| = 5.66$$

$a_t =$ _____

$a_n =$ _____ $= 11.31$ m/s^2

$a = \sqrt{(4)^2 + (11.31)^2} = 12.0$ m/s^2 *Ans.*

12-19. When the motorcyclist is at A he increases his speed along the vertical circular parth at the rate of $\dot{v} = (0.3t)$ ft/s^2, where t is in seconds. If he starts from rest when he is at A, determine his velocity and acceleration when he reaches B.

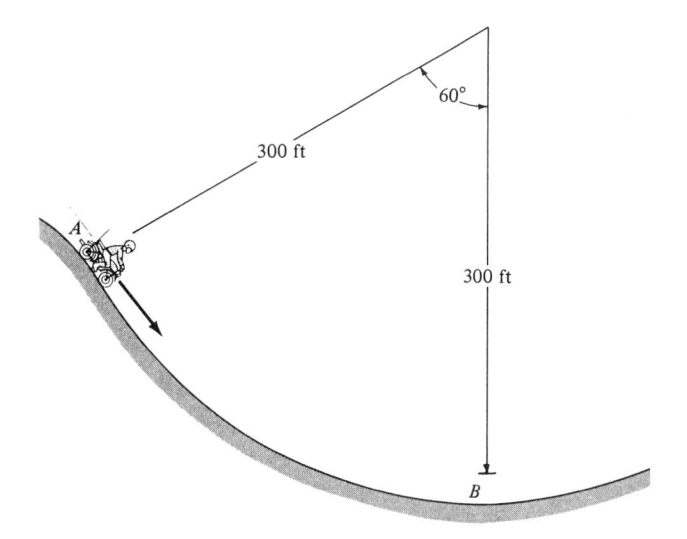

Solution

Since $v = 0$ at $t = 0$, v as a function of time is

_____

Since $s = 0$ at $t = 0$, s as a function of time is

_____

Length of path is

$$s_{AB} = \underline{\hspace{4cm}} = 314.16 \text{ ft}$$

Time to reach B is

$$t_B = \underline{\hspace{4cm}} = 18.45 \text{ s}$$

Thus

$$v|_{t_B} = \underline{\hspace{4cm}} = 51.1 \text{ ft/s} \qquad Ans.$$

$$a_t|_{t_B} = \underline{\hspace{4cm}} = 5.54 \text{ ft/s}^2$$

$$a_n|_{t_B} = \underline{\hspace{4cm}} = 8.70 \text{ ft/s}^2$$

$$a = \sqrt{(5.54)^2 + (8.70)^2} = 10.3 \text{ ft/s}^2 \qquad Ans.$$

20 Study Guide and Problems

12-20. A package is dropped from the plane which is flying with a constant horizontal velocity of $v_A = 150$ ft/s. Determine the tangential and normal components of acceleration and the radius of curvature of the path of motion (a) at the moment the package is released at A, where it has a horizontal velocity of $v_A = 150$ ft/s and $h = 1500$ ft, and (b) *just before* it strikes the ground at B.

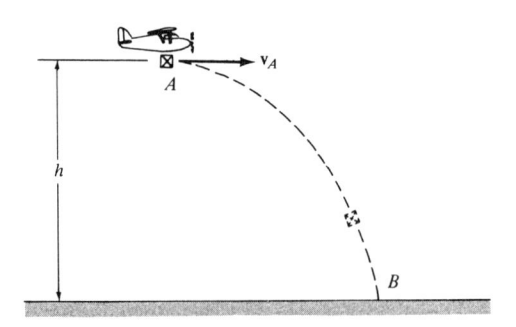

Solution:

Since the acceleration is simply 32.2 ft/s² ↓, then

a) $(a_n)_A =$ _____

$(a_t)_A =$ _____ Ans.

Knowing v_A and $(a_n)_A$, then

$\rho_A =$ _____ $= 699$ ft Ans.

b) $(v_B)_x =$ _____

To determine $(v_B)_y$, apply

$+\downarrow \ (v_B)_y^2 = (v_A)_y^2 + 2a_c[(s_B)_y - (s_A)_y]$

_____

$(v_B)_y = 310.8$

$v_B = \sqrt{(150)^2 + (310.8)^2} = 345$ ft/s

$\theta = \tan^{-1}\dfrac{310.8}{150} = 64.23°$

$(a_n)_B =$ _____ $= 14.0$ ft/s² Ans.

$(a_t)_B =$ _____ $= 29.0$ ft/s² Ans.

$\rho_B =$ _____ $= 8.51(10^3)$ ft Ans.

Curvilinear Motion: Cylindrical Components

12-21. A car is traveling along the circular curve of radius $r = 300$ ft. At the instant shown, its angular rate of rotation is $\dot{\theta} = 0.4$ rad/s, which is increasing at the rate of $\ddot{\theta} = 0.2$ rad/s². Determine the magnitudes of the car's velocity and acceleration at this instant.

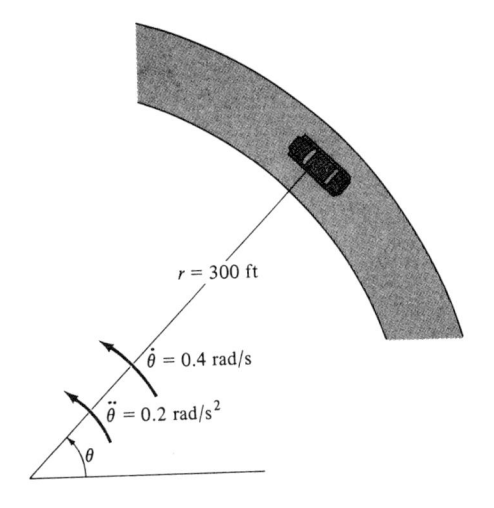

Solution

$\dot{\theta} = 0.4$ rad/s

$\ddot{\theta} = $ _____

$r = 300$ ft

$\dot{r} = $ _____

$\ddot{r} = $ _____

$v_r = \dot{r} = 0$

$v_\theta = r\dot{\theta} = 300(0.4) = 120$ ft/s

$v = \sqrt{(0)^2 + (120)^2} = 120$ ft/s Ans.

$a_r = \ddot{r} - r\dot{\theta}^2 = 0 - 300(0.4)^2 = -48$ ft/s²

$a_\theta = r\ddot{\theta} + 2\dot{r}\dot{\theta} = 300(0.2) + 2(0)(0.4) = 60$ ft/s²

$a = \sqrt{(-48)^2 + (60)^2} = 76.8$ ft/s² Ans.

12-22. For a short time the position of a roller-coaster car along its path is defined by the equations $r = 25$ m, $\theta = (0.3t)$ rad, and $z = (-8\cos\theta)$ m, where t is measured in seconds. Determine the magnitudes of the car's velocity and acceleration when $t = 4$ s.

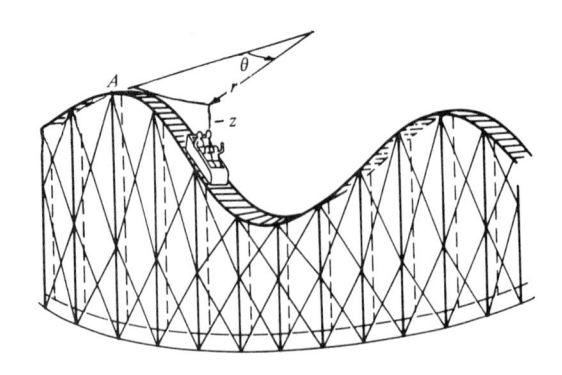

Solution

$r = $ _____

$\dot{r} = $ _____

$\ddot{r} = $ _____

$\theta = $ _____

$\dot{\theta} = $ _____

$\ddot{\theta} = $ _____

$z = $ _____

$\dot{z} = $ _____

$\ddot{z} = $ _____

$v_r = \dot{r} = 0$

$v_\theta = r\dot{\theta} = 25(0.3) = 7.5$ m/s

$v_z = \dot{z} = 2.4 \sin(0.3(4)) = 2.23$ m/s

$v = \sqrt{(7.5)^2 + (2.23)^2} = 7.82$ m/s²   Ans.

$a_r = \ddot{r} - r\dot{\theta}^2 = 0 - 25(0.3)^2 = 2.25$ m/s²

$a_\theta = r\ddot{\theta} + 2\dot{r}\dot{\theta} = 0 + 0 = 0$

$a_z = \ddot{z} = 0.72 \cos(0.3(4)) = 0.261$ m/s²

$a = \sqrt{(-2.25)^2 + (0.261)^2} = 2.26$ m/s²   Ans.

12-23. The slotted link is pinned at O, and as a result of rotation it drives the peg P along the horizontal guide. Compute the magnitudes of the velocity and acceleration of P as a function of θ if $\theta = (3t)$ rad, where t is measured in seconds.

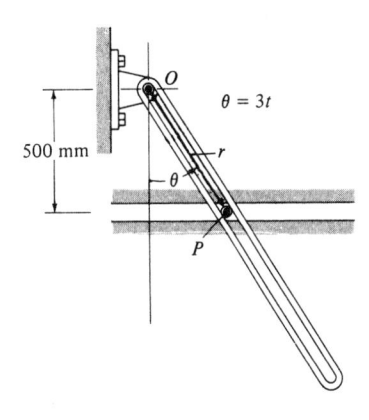

Solution

In the following use the identity $1 + \tan^2\theta = \sec^2\theta$

$\theta =$ _____

$\dot\theta =$ _____

$\ddot\theta =$ _____

$r =$ _____ $= 500 \sec\theta$

$\dot r =$ _____ $= 1500 \sec\theta \tan\theta$

$\ddot r =$ _____ $= 4500 (\sec\theta)(2\tan^2\theta + 1)$

$v_r = \dot r = 1500 \sec\theta \tan\theta$

$v_\theta = r\dot\theta = 1500 \sec\theta$

$v = \sqrt{(1500)^2 \sec^2\theta \tan^2\theta + (1500)^2 \sec^2\theta}$

$v = 1500 \sec^2\theta$ mm/s *Ans.*

$a_r = \ddot r - r\dot\theta^2 = 4500 \sec\theta (2\tan^2\theta + 1) - 500 \sec\theta (3)^2 = 9000 \sec\theta \tan^2\theta$

$a_\theta = r\ddot\theta + 2\dot r\dot\theta = 0 + 2(1500 \sec\theta \tan\theta)(3) = 9000 \sec\theta \tan\theta$

$a = 9000 \sqrt{\sec^2\theta \tan^4\theta + \sec^2\theta \tan^2\theta}$

$a = 9000 \sec^2\theta \tan\theta$ mm/s² *Ans.*

24 Study Guide and Problems

12-24. The cylindrical cam C is held fixed while the rod AB and bearings E and F rotate about the vertical axis of the cam at a constant rate of $\dot{\theta} = 4$ rad/s. If the rod is free to slide through the bearings, determine the magnitudes of the velocity and acceleration of the guide D on the rod as a function of θ. The guide follows the groove in the cam, and the groove is defined by the equations $r = 0.25$ ft and $z = (0.25 \cos\theta)$ ft.

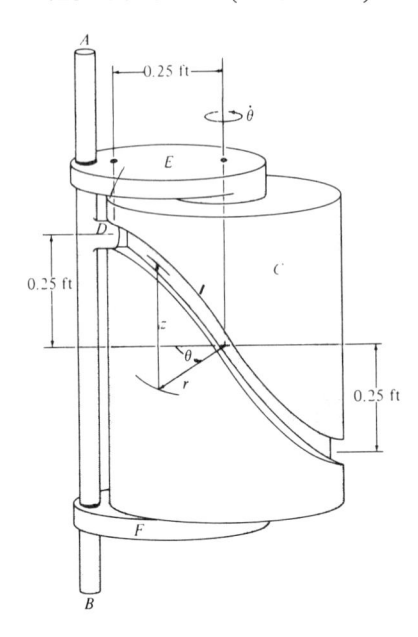

Solution

$r = $ _____

$\dot{r} = $ _____

$\ddot{r} = $ _____

$\dot{\theta} = $ _____

$\ddot{\theta} = $ _____

$z = $ _____

$\dot{z} = $ _____ $= -\sin\theta$

$\ddot{z} = $ _____ $= -4\cos\theta$

$v_r = \dot{r} = 0$

$v_\theta = r\dot{\theta} = (0.25)(4) = 1$

$v_z = \dot{z} = -\sin\theta$

$v = \sqrt{1 + \sin^2\theta}$ ft/s **Ans.**

$a_r = \ddot{r} - r\dot{\theta}^2 = 0 - 0.25(4)^2 = -4$ ft/s

$a_\theta = r\ddot{\theta} + 2\dot{r}\dot{\theta} = 0 + 0 = 0$

$a_z = \ddot{z} = -4\cos\theta$

$a = 4\sqrt{1 + \cos^2\theta}$ ft/s² **Ans.**

12-25. A double collar C is pin-connected together such that one collar slides over a fixed rod and the other slides over a rotating rod. If the geometry of the fixed rod for a short distance can be defined by a lemniscate, $r^2 = (4 \cos 2\theta)$ ft^2, determine the collar's radial and transverse components of velocity and acceleration at the instant $\theta = 0°$ as shown. Rod OA is rotating at a constant rate of $\dot\theta = 6$ rad/s.

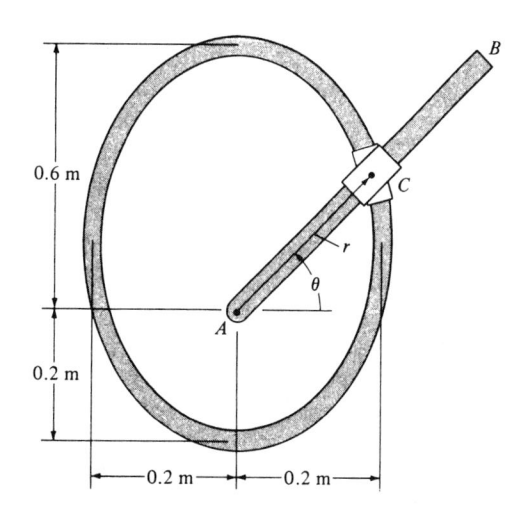

Solution

$$r^2 = 4\cos 2\theta$$

Taking two successive time derivatives yields

$r\dot r = $ _____

$r\ddot r + \dot r^2 = $ _____

$\theta = 0$

$\dot\theta = 6$

$\ddot\theta = $ _____

At $\theta = 0°$,

$r = 2$ ft

$\dot r = 0$

$\ddot r = -144$ ft/s^2

$v_r = \dot r = 0$ Ans.

$v_\theta = r\dot\theta = 2(6) = 12$ ft/s   Ans.

$a_r = \ddot r - r\dot\theta^2 = -144 - 2(6)^2 = -216$ ft/s^2 Ans.

$a_\theta = r\ddot\theta + 2\dot r\dot\theta = 2(0) + 2(0)(6) = 0$ Ans.

Absolute-Dependent-Motion Analysis of Two Particles

12-26. If the end of the cable at A is pulled down with a speed of 2 m/s, determine the speed at which block B arises.

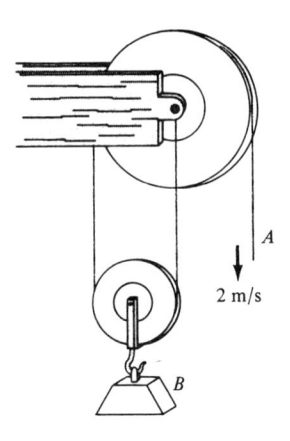

Solution

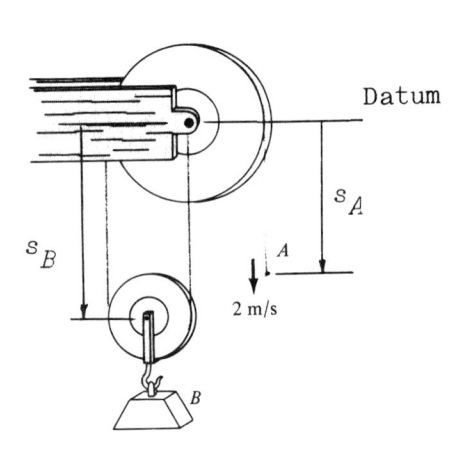

Position Equation

_____

Velocity

$v_B = $ _____

$v_B = 1$ m/s ↑

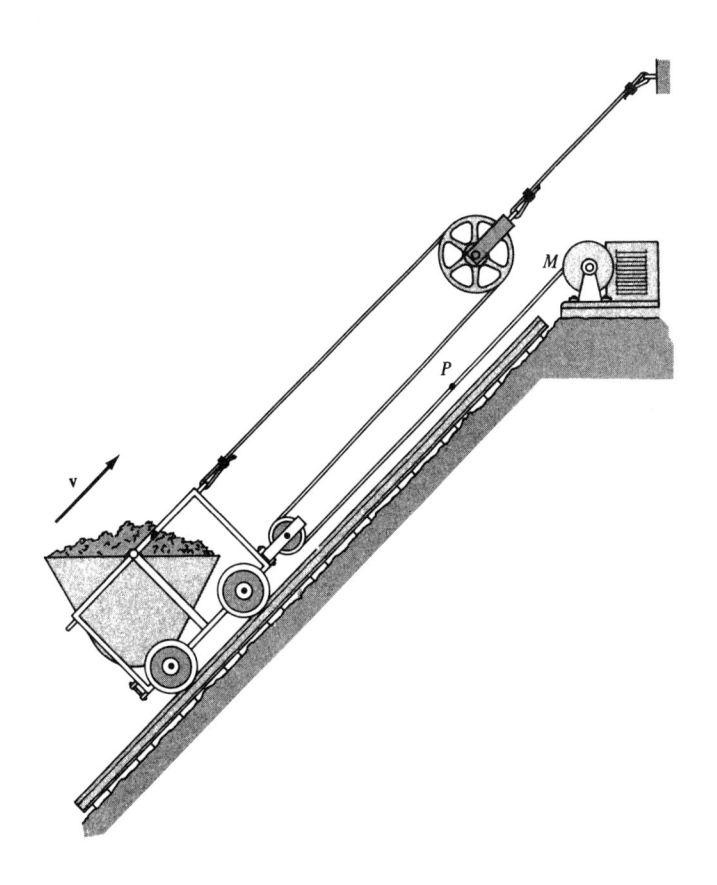

12-27. The mine car is being pulled up the inclined plane using the motor M and the rope-and-pulley arrangement shown. Determine the speed v_P at which a point P on the cable must be traveling toward the motor to move the car up the plane with a constant speed of $v = 5$ m/s.

Solution

Establish position coordinates, measured from the pulley (fixed point) to the mine car and to P.

Position equation

Velocity

$v_P = 15$ m/s   Ans.

12-28. The block B is suspended from a cable that is attached to the block at E, wraps around three pulleys, and is tied to the back of a truck. If the truck starts from rest when x_D is zero, and moves forward with a constant acceleration of $a_D = 2$ m/s², determine the speed of the block at the instant $x_D = 3$ m. Neglect the size of the pulleys in the calculation. When $x_D = 0$, $x_C = 5$ m so that points C and D are at the same elevation. *Hint*: Relate the coordinates x_C and x_D using the problem geometry, then take the time derivative.

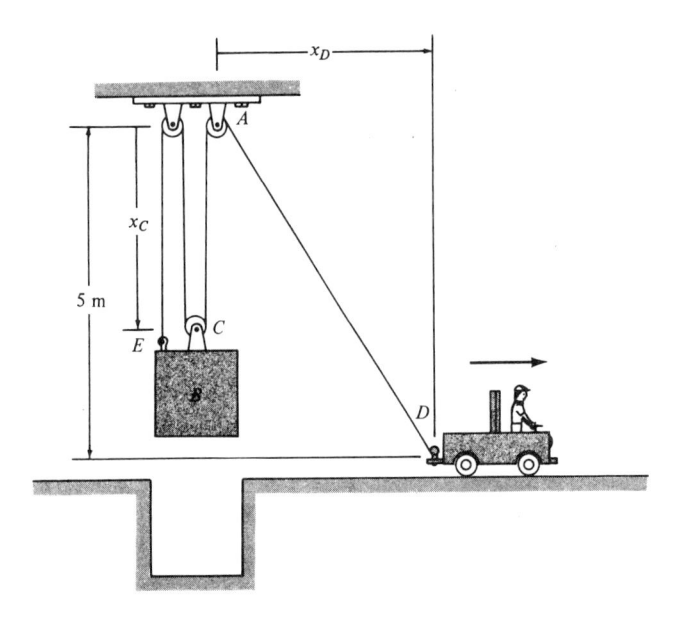

Solution

When $x_D = 0$, the total cord length is

$$l = \underline{\hspace{4cm}}$$

When $x_D > 0$, express the cord length in terms of x_D and x_C.

$$l = \underline{\hspace{4cm}}$$

Taking the time derivative,

$$\underline{\hspace{10cm}} \quad (1)$$

Since $x_D = 3$m, also

$$\overset{+}{\rightarrow} (v_D)_2^2 = (v_D)_1^2 + 2a_D[(x_D)_2 - (x_D)_1]$$

$$(v_D)_2^2 = \underline{\hspace{4cm}}$$

$$(v_D)_2 = \dot{x}_D = 3.46 \text{ m/s}$$

Eqation (1) becomes

$$3v_C + \frac{1}{2}\{(3)^2 + (5)^2\}^{-1/2} 2(3)(3.46) = 0$$

$$v_C = -0.594 \text{ m/s} = 0.594 \text{ m/s}\uparrow \qquad \textit{Ans.}$$

Relative-Motion Analysis of Two Particles Using Translating Axes

12-29. A fly traveling horizontally at a constant speed enters the open window of a train and leaves through the opposite window 3 m away 0.75 s later. If the fly travels perpendicular to the train's motion and the train is traveling at 3 m/s, determine the speed and direction of flight of the fly observed by a passenger on the train.

Solution

Compute the velocity of the fly:

$$v_f = \underline{\hspace{4cm}} = 4 \text{ m/s} \rightarrow$$

Apply the relative velocity equation

$$\mathbf{v}_f = \mathbf{v}_T + \mathbf{v}_{f/T}$$

_____

$(v_{f/T})_x = 4$ m/s $\rightarrow$

$(v_{f/T})_y = 3$ m/s $\downarrow$

$v_{f/T} = \sqrt{(4)^2 + (3)^2} = 5$ m/s

$\theta = \tan^{-1} \dfrac{3}{4} = 36.9°$

12-30. At the instant shown, cars A and B are traveling at speeds of 20 mi/h and 45 mi/h, respectively. If B is accelerating at 1600 mi/h² while A maintains a constant speed, determine the velocity and acceleration of A with respect to B.

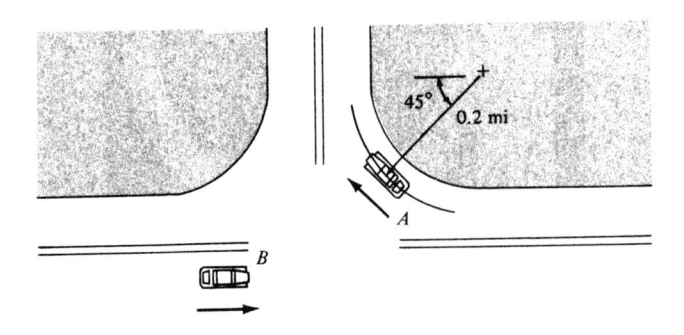

Solution

Apply the relative velocity equation

$$\mathbf{v}_A = \mathbf{v}_B + \mathbf{v}_{A/B}$$

$\xrightarrow{+}$ $-20\cos 45° =$ _____

$+\uparrow$ $20\sin 45° =$ _____

$(v_{A/B})_x = -59.14 = 59.14$ mi/h $\leftarrow$

$(v_{A/B})_y = 14.14$ mi/h $\uparrow$

$v_{A/B} = \sqrt{(59.14)^2 + (14.14)^2} = 60.8$ mi/h Ans.

$\theta = \tan^{-1} \dfrac{14.14}{59.14} = 13.5°$ Ans.

Apply the relative acceleration equation

$$\mathbf{a}_A = \mathbf{a}_B + \mathbf{a}_{A/B}$$

$\xrightarrow{+}$ $1414.2 = 1600 + (a_{A/B})_x$

$+\uparrow$ $1414.2 = 0 + (a_{A/B})_y$

$(a_{A/B})_x = -185.8 = 185.8$ mi/h² $\leftarrow$

$(a_{A/B})_y = 1414.2$ mi/h² $\uparrow$

$a_{A/B} = \sqrt{(185.8)^2 + (1414.2)^2} =$ Ans.

$\phi = \tan^{-1}\left(\dfrac{1414.2}{185.8}\right) = 82.5°$

12-31. A passenger in the automobile B observes the motion of the train car A. At the instant shown, the train has a speed of 18 m/s and is reducing its speed at a rate of 1.5 m/s². The automobile is accelerating at 2 m/s² and has a speed of 25 m/s. Determine the velocity and acceleration of A with respect to B. The train is moving along a curve of radius $r = 300$ m.

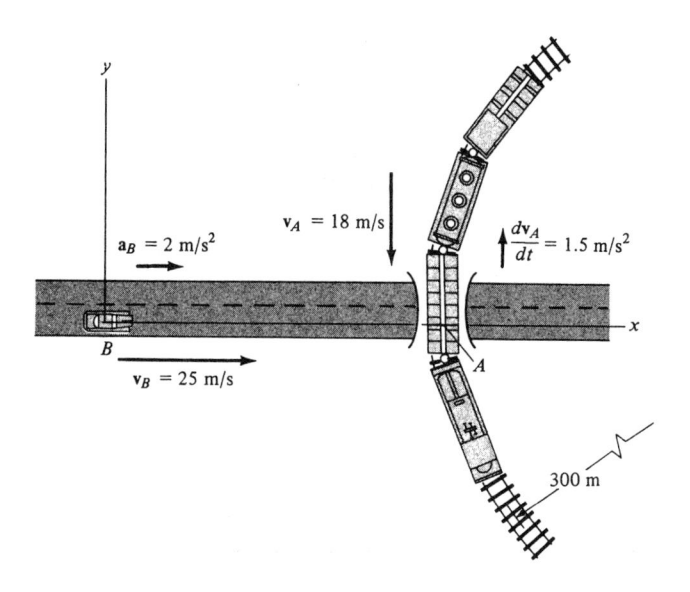

Solution

Apply the relative velocity equation

$$\mathbf{v}_A = \mathbf{v}_B + \mathbf{v}_{A/B}$$

$(v_{A/B})_x = 25$ m/s ←

$(v_{A/B})_y = 18$ m/s ↓

$v_{A/B} = 30.8$ m/s   Ans.

Apply the relative acceleration equation

$$\mathbf{a}_A = \mathbf{a}_B + \mathbf{a}_{A/B}$$

$(a_{A/B})_x = 0.920$ m/s² ←

$(a_{A/B})_y = 1.50$ m/s² ↑

$a_{A/B} = 1.76$ m/s²   Ans.

12-32. If the hoist H is moving upward at 6 ft/s, determine the speed at which the motor M must draw in the supporting cable.

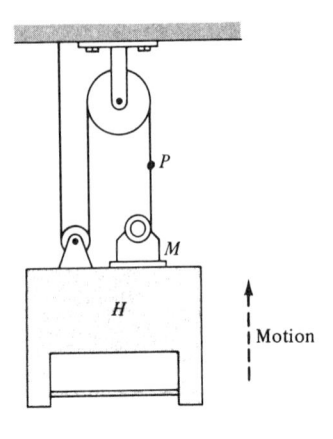

Solution

Establish datum through top pulley. Specify position of P and pulley on hoist H.

Position equation

Velocity equation

$$v_P = 12 \text{ ft/s} \downarrow$$

Apply the relative velocity equation

$$\mathbf{v}_P = \mathbf{v}_H + \mathbf{v}_{P/H}$$

$$v_{P/H} = 18 \text{ ft/s} \downarrow \qquad Ans.$$

12-33. The pilot of fighter plane F is following 1.5 km behind the pilot of bomber B. Both planes are originally traveling at 120 m/s. In an effort to pass the bomber, the pilot in F gives his plane a constant acceleration of 12 m/s². Determine the speed at which the pilot in the bomber sees the pilot of the fighter plane pass if at the start of the passing operation the bomber is decelerating at 3 m/s². Neglect the effect of any turning.

Solution

At any instant t, the fighter plane has a position

$$\xrightarrow{+} s_F = (s_F)_1 + (v_F)_1 t + \frac{1}{2} a_1 t^2$$

$s_F = $ _____

The bomber has a position

$$\xrightarrow{+} s_B = (s_B)_1 + (v_B)_1 t + \frac{1}{2} a_B t^2$$

$s_B = $ _____

Require

$$s_B = s_F$$

gives

$$1500 + 120t - 1.5t^2 = 120t + 6t^2$$
$$t = 14.14 \text{ s}$$

Velocity of fighter is

$$v_F = (v_F)_1 + a_F t$$

$v_F = $ _____ = 289.7 m/s

Velocity of bomber is

$$v_B = (v_B)_1 + a_B t$$

$v_B = $ _____ = 77.6 m/s

Thus

$$\mathbf{v}_F = \mathbf{v}_B + \mathbf{v}_{F/B}$$

$$\xrightarrow{+} 289.7 = 77.6 + v_{F/B}$$

$$v_{F/B} = 212 \text{ m/s} \rightarrow \qquad \text{Ans.}$$

13 Kinetics of a Particle: Force And Acceleration

Equations of Motion: Rectangular Coordinates

13-1. Each of the three barges has a mass of 30 Mg, whereas the tugboat has a mass of 12 Mg. As the barges are being pulled forward with a constant velocity of 4 m/s, the tugboat must overcome the frictional resistance of the water, which is 2 kN for each barge and 1.5 kN for the tugboat. If the cable between A and B breaks, determine the acceleration of the tugboat.

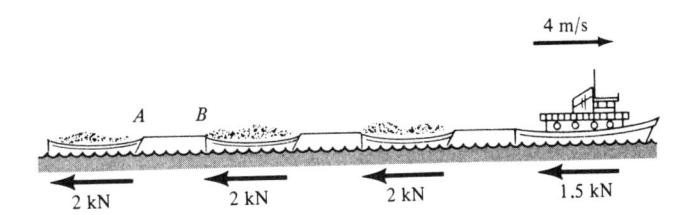

Solution

When cable breaks

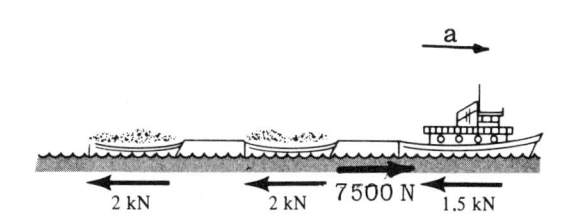

$\xrightarrow{+} \Sigma F_x = ma_x;$ _____

$a = 0.0278 \text{ m/s}^2 \rightarrow$ *Ans.*

13-2. A block having a mass of 2 kg is placed on a spring scale located in an elevator that is moving downward. If the scale reading, which measures the force in the spring, is 20 N, determine the acceleration of the elevator. Neglect the mass of the scale.

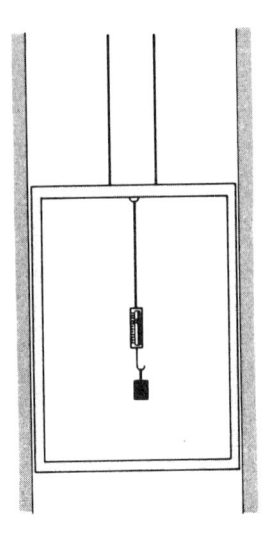

Solution

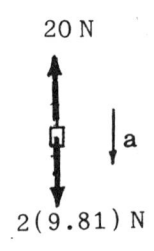

$+\downarrow \Sigma F_y = ma_y;$ _____

$a = 0.190 \text{ m/s}^2 \uparrow$ *Ans.*

13-3. The 300-kg bar B, originally at rest, is being towed over a series of small rollers. Compute the force in the cable when $t = 5$ s, if the motor M is drawing in the cable for a short time at a rate of $v = (0.4t^2)$ m/s, where t is in seconds ($0 \leq t \leq 6$ s). How far does the bar move in 5 s? Neglect the mass of the cable, pulley P, and the rollers.

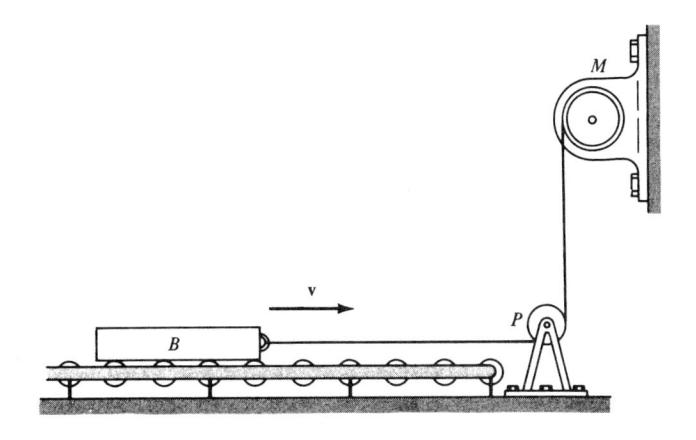

Solution

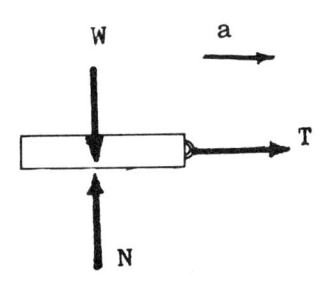

Since $v = 0.4t^2$; then a as a function of time is

$$a = \underline{\hspace{4cm}}$$

at $t = 5$ s,

$$a = 4 \text{ m/s}^2$$

$\xrightarrow{+} \Sigma F_x = ma_x;$ \underline{\hspace{4cm}}

$$T = 1200 \text{ N}$$ Ans.

$$ds = v \, dt$$

$$\int_0^s ds = \int_0^s 0.4 t^2 \, dt$$

$$s = \frac{0.4}{3}(5)^3$$

$$s = 16.7 \text{ m}$$ Ans.

13-4. A 1.5-lb brick is released from rest at A and slides down the inclined roof. If the coefficient of friction between the roof and the brick is $\mu = 0.3$, determine the speed at which the brick strikes the gutter G.

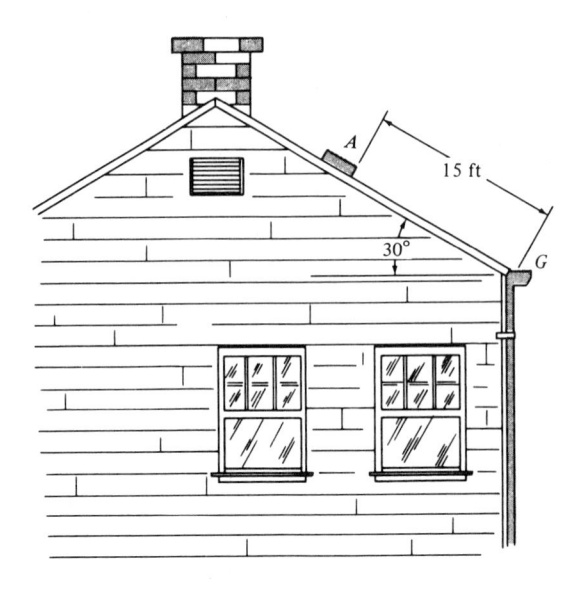

Solution

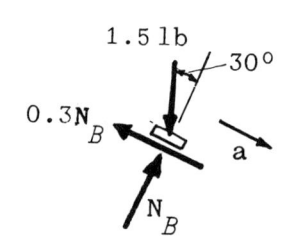

$+\nwarrow \Sigma F_y = ma_y$ _____

$$N_B = 1.30 \text{ lb}$$

$\overset{+}{\searrow} \Sigma F_x = ma_x;$ _____

$$a = 7.728 \text{ ft/s}^2$$

$\overset{+}{\searrow} v_2^2 = v_1^2 + 2a_c s$ _____

$$v_2 = 15.2 \text{ ft/s} \qquad \qquad Ans.$$

13-5. The 2-kg shaft *CA* passes through a smooth journal bearing at *B*. Initially, the springs, which are coiled loosely around the shaft, are unstretched when no force is applied to the shaft. In this position $s = s' = 250$ mm and the shaft is originally at rest. If a horizontal force of $F = 5$ kN is applied, determine the speed of the shaft at the instant $s = 50$ mm, $s' = 450$ mm. The ends of the springs are attached to the bearing at *B* and the caps at *C* and *A*.

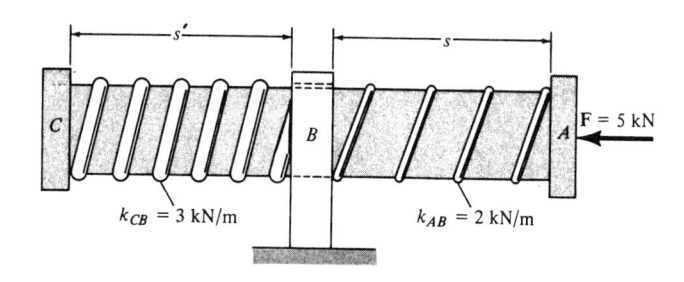

Solution

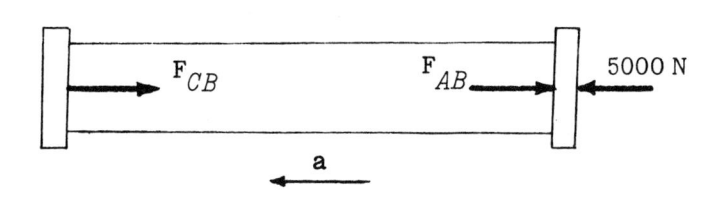

$F_{CB} = k_{CB} x = 3000x$

$F_{AB} = k_{AB} x = 2000x$

$\overset{+}{\leftarrow} \Sigma F_x = ma_x;$ _____

$2500 - 2500x = a$

Kinematics: $a\,dx = v\,dv$, thus

$$\int_0^{0.2} \underline{\hspace{5cm}} = \int_0^v v\,dv$$

$$2500(0.2) - \frac{2500(0.2)^2}{2} = \frac{v^2}{2}$$

$v = 30$ m/s *Ans.*

13-6. Determine the acceleration of block A when the system is released. The coefficient of friction and the weight of each block are indicated in the figure. Neglect the mass of the pulleys and cords.

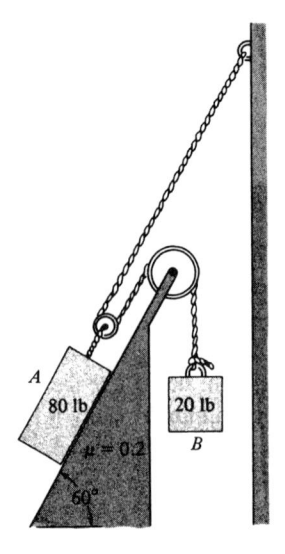

Solution

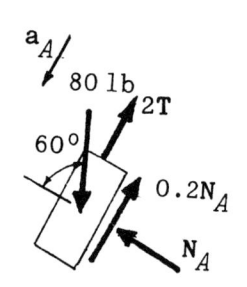

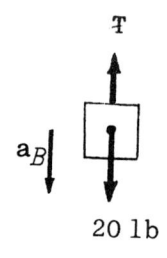

Block A:

$+\nwarrow \Sigma F_y = ma_y;$ _____

$+\swarrow \Sigma F_x = ma_x;$ _____

Block B:

$+\downarrow \Sigma F_y = ma_y;$ _____

Show on the figure that

$$2a_A = -a_B$$

Note that the position coordinates for s_A and s_B, and a_A and a_B must be in the *same* directions
Solving,

$$N_A = 40 \text{ lb}, \ T = 25.3 \text{ lb}$$
$$a_B = 8.57 \text{ ft/s}^2 \uparrow \quad a_A = 4.28 \text{ ft/s}^2 \swarrow \qquad \qquad Ans.$$

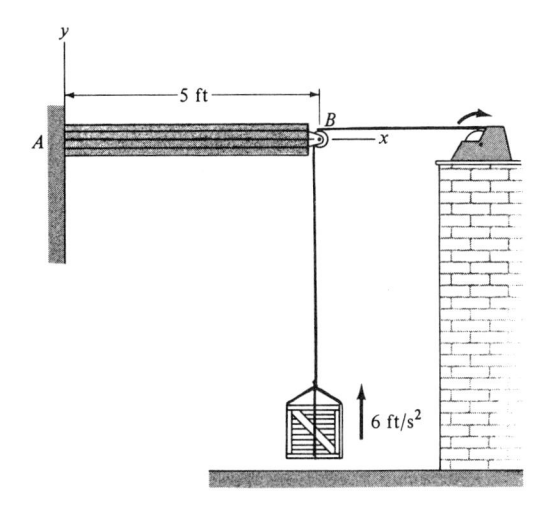

13-7. The 30-lb crate is being hoisted upward with a constant acceleration of 6 ft/s². If the uniform beam AB has a weight of 200 lb, determine the components of reaction at A. Neglect the size and mass of the pulley at B. *Hint*: First find the tension in the cable, then analyze the forces on the beam using statics.

Solution

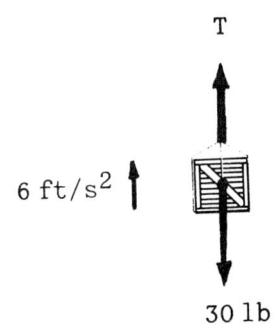

$+\uparrow \Sigma F_y = ma_y;$ _____

$$T = 35.59 \text{ lb}$$

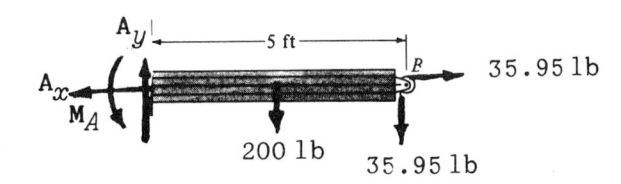

$\xrightarrow{+} \Sigma F_x = 0;$ _____

$$A_x = 35.6 \text{ lb} \qquad \qquad Ans.$$

$+\uparrow \Sigma F_y = 0;$ _____

$$A_y = 236 \text{ lb} \qquad \qquad Ans.$$

$\left(+ \Sigma M_A = 0;\right.$ _____

$$M_A = 678 \text{ lb} \cdot \text{ft} \qquad \qquad Ans.$$

Equations of Motion: Normal and Tangential Coordinates

13-8. A boy twirls a 15-lb bucket of water in a vertical circle. If the radius of curvature of the path is 4 ft, determine the minimum speed the bucket must have when it is overhead at A so no water spills out. Neglect the size of the bucket in the calculation. If the bucket were moving at a slightly slower rate than that calculated, would the water fall on the boy when it starts to spill out at A? Explain.

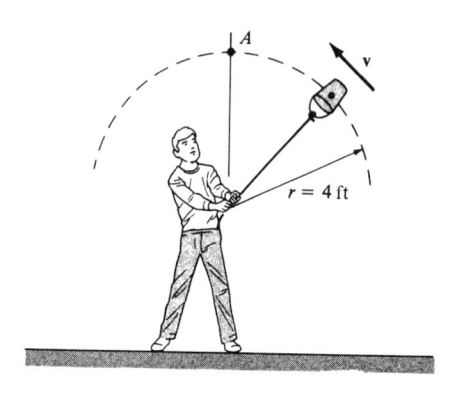

Solution

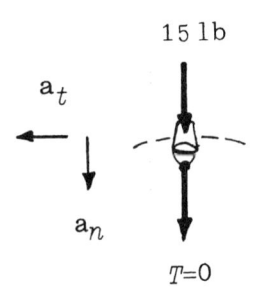

$+\downarrow \Sigma F_n = ma_n;$ _____

$$v = 11.3 \text{ ft/s} \quad \text{Ans.}$$

Will water spill on the boy?

No/Yes

Explain _____

13-9. A toboggan and rider have a total mass of 100 kg and travel down along the (smooth) slope defined by the equation $y = 0.2x^2$. At the instant $x = 8$ m, the toboggan's speed is 4 m/s. At this point, determine the rate of increase in speed and the normal force which the toboggan exerts on the slope. Neglect the size of the toboggan and rider for the calculation.

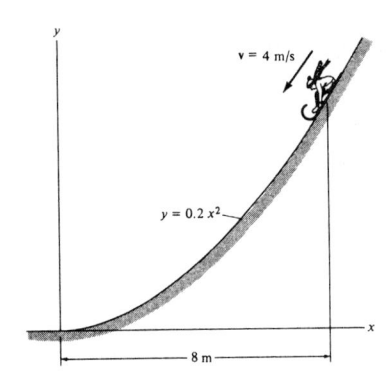

Solution

$$y = 0.2x^2$$

$$\left.\frac{dy}{dx}\right|_{x=8} = 0.4x = 3.2, \quad \theta = \tan^{-1} 3.2 = 72.6°$$

$$\frac{d^2y}{dx^2} = 0.4$$

$$\rho = \left|\frac{\left[1 + \left(\frac{dy}{dx}\right)^2\right]^{3/2}}{\left(\frac{d^2y}{dx^2}\right)}\right| = \left|\frac{[1 + (3.2)^2]^{3/2}}{0.4}\right|$$

$$\rho = 94.2 \text{ m}$$

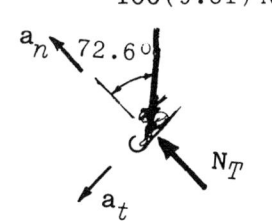

$\nwarrow + \Sigma F_n = ma_n;$

$$N_T = 310 \text{ N} \qquad \qquad Ans.$$

$\stackrel{+}{\swarrow} \Sigma F_t = ma_t;$

$$a_t = 9.36 \text{ m/s}^2 \qquad \qquad Ans.$$

13-10. The pendulum bob B has a weight of 5 lb and is released from rest in the position shown, $\theta = 0°$. Determine the tension in string BC just after the bob is released, $\theta = 0°$, and also at the instant the bob reaches point D, $\theta = 45°$.

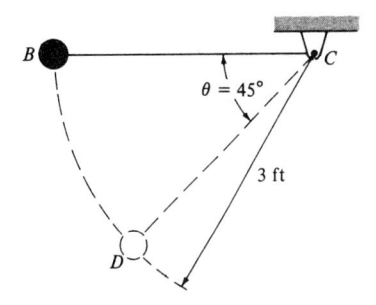

Solution

Since the initial velocity is zero

$$a_n = 0$$

$\xrightarrow{+} \Sigma F_n = ma_n, \quad T = 0 \quad$ *Ans.*

$+\searrow \Sigma F_t = ma_t;$ _____

$$a_t = 32.2 \cos\theta \qquad (1)$$

$\overset{+}{\nearrow} \Sigma F_n = ma_n;$ _____ (2)

$$v\,dv = a_t\,ds = a_t(r\,d\theta) = a_t(3\,d\theta)$$

Using Eq. (1),

$$\int_0^v v\,dv = 96.6 \int_0^{45°} \cos\theta\,d\theta$$

$$\frac{1}{2}v^2 = 96.6\,[\sin\theta]_0^{45°} = 68.3$$

$$v = 11.7 \text{ ft/s}$$

Substituting into Eq. (2) with $\theta = 45°$ yields

$$T = 10.6 \text{ lb} \qquad \qquad \textit{Ans.}$$

13-11. A ball having a mass of 2 kg rolls within a vertical circular slot. If it is released from rest when $\theta = 10°$, determine the force it exerts on the slot when it arrives at points A and B. Neglect the rolling motion of the ball in the calculation.

Solution

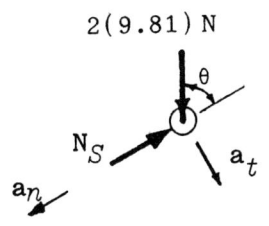

$+ \searrow \Sigma F_t = ma_t;$

$$a_t = 9.81 \sin\theta \quad (1)$$

$+ \swarrow \Sigma F_n = ma_n; \quad (2)$

$$v\,dv = a_t\,ds = a_t(0.8\,d\theta)$$

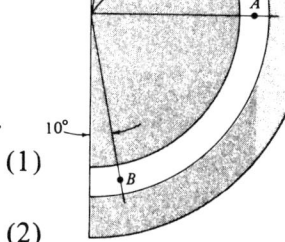

Using Eq. (1):

$$v\,dv = 9.81 \sin\theta\,(0.8\,d\theta)$$

At A;

$$\int_0^{v_A} v\,dv = 7.848 \int_{10°}^{90°} \sin\theta\,d\theta$$

$$\frac{1}{2} v_A^2 = 7.848[-\cos 90° + \cos 10°]$$

$$v_A = 3.93 \text{ m/s}$$

At B;

$$\int_0^{v_B} v\,dv = 7.848 \int_{10°}^{170°} \sin\theta\,d\theta$$

$$\frac{v_B^2}{2} = 7.848[-\cos 170° + \cos 10°]$$

$$v_B = 5.56 \text{ m/s}$$

Using Eq. (2)

At A:

$$\theta = 90°; \quad N_s = -2\left(\frac{(3.93)^2}{0.8}\right) + 0$$

$$N_s = -38.6 \text{ N} \qquad \textit{Ans.}$$

At B:

$$\theta = 170°; \quad N_s = 2(9.81)\cos 170° - 2\left(\frac{(5.56)^2}{0.8}\right)$$

$$N_s = -96.6 \text{ N} \qquad \textit{Ans.}$$

Equations of Motion: Cylindrical Coordinates

13-12. A particle, having a mass of 1.5 kg, moves along a three-dimensional path defined by the equations $r = (4 + 3t)$ m, $\theta = (t^2 + 2)$ rad, and $z = (6 - t^3)$ m, where t is in seconds. Determine the r, θ, and z components of force which the path exerts on the particle when $t = 2$ s.

Solution:

At $t = 2$ s

$r = $ _____

$\dot{r} = $ _____

$\ddot{r} = $ _____

$\theta = $ _____

$\dot{\theta} = $ _____

$\ddot{\theta} = $ _____

$z = $ _____

$\dot{z} = $ _____

$\ddot{z} = $ _____

$a_r = \ddot{r} - r(\dot{\theta})^2 = 0 - 10(4)^2 = -160$

$a_\theta = r\ddot{\theta} + 2\dot{r}\dot{\theta} = 10(2) + 2(3)(4) = 44$

$a_z = \ddot{z} = -12$

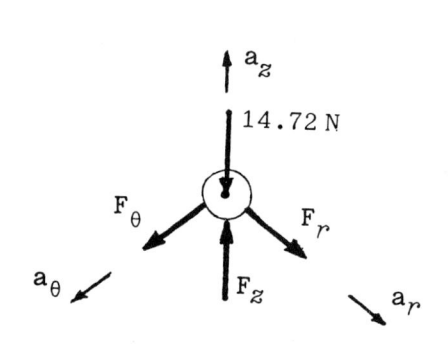

$\Sigma F_r = ma_r; \quad F_r = 1.5(-160); \quad F_r = -240$ N *Ans.*

$\Sigma F_\theta = ma_\theta; \quad F_\theta = 1.5(44); \quad F_\theta = 66$ N *Ans.*

$\Sigma F_z = ma_z; \quad F_z - 14.72 = 1.5(-12); \quad F_z = -3.28$ N *Ans.*

13-13. A smooth can C, having a mass of 2 kg, is lifted from a feed at A to a ramp at B by a forked rotating rod. If the rod maintains a constant angular motion of $\dot{\theta} = 0.5$ rad/s, determine the force which the rod exerts on the can at the instant $\theta = 30°$. Neglect the effects of friction in the calculation. The ramp from A to B is circular, having a radius of 700 mm.

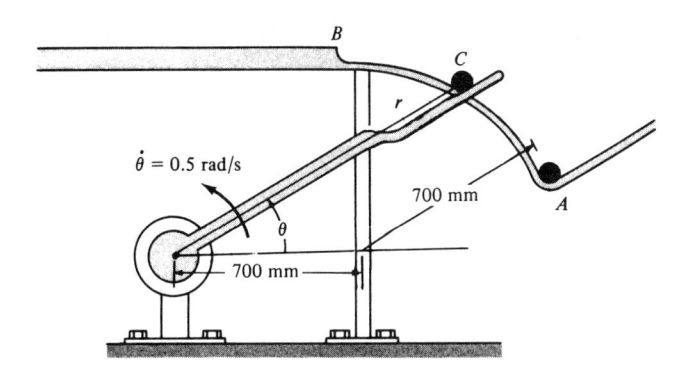

Solution

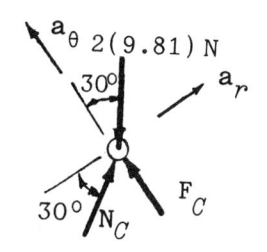

$+\nearrow \Sigma F_r = ma_r;$ _____ $= 2a_r$ (1)

$+\nwarrow \Sigma F_\theta = ma_\theta;$ _____ $= 2a_\theta$ (2)

$\dot{\theta} = 0.5$

$\ddot{\theta} = $ _____

$r = 2(0.7) \cos\theta = 1.4 \cos\theta$

$\dot{r} = $ _____ $= -0.7 \sin\theta$

$\ddot{r} = $ _____ $= -0.35 \cos\theta$

At $\theta = 30°$,

$$a_r = \ddot{r} - r(\dot{\theta})^2 = -0.35 \cos 30° - (1.4 \cos 30°)(0.5)^2 = -0.6062$$

$$a_\theta = r\ddot{\theta} + 2\dot{r}\dot{\theta} = 0 + 2(-0.7 \sin 30°)(0.5) = -0.350$$

Substituting into Eqs. (1) and (2) and solving

$$F_C = 11.3 \text{ N} \qquad \text{Ans.}$$
$$N_C = 9.93 \text{ N}$$

48 Study Guide and Problems

13-14. The spool, which has a weight of 2 lb, slides along the smooth *horizontal* spiral rod, $r = (2\theta)$ ft, where θ is in radians. If its angular rate of rotation is constant and equals $\dot{\theta} = 4$ rad/s, determine the tangential force **P** needed to cause the motion and the normal force that the spool exerts on the rod at the instant $\theta = 90°$.

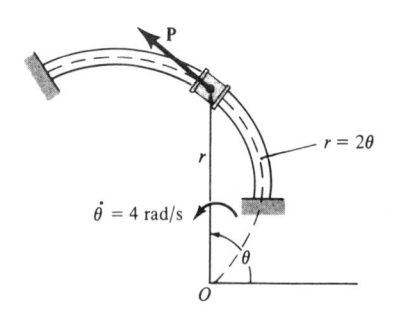

Solution

$$\tan \psi = r / \frac{dr}{d\theta} = \underline{\hspace{2cm}} \bigg|_{90°} = \pi/2$$

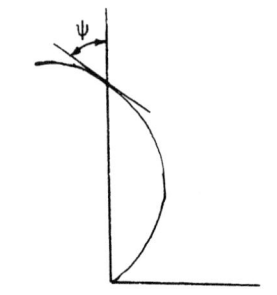

$\psi = 57.52°$

$+\uparrow \Sigma F_r = ma_r;$ _____ $= \dfrac{2}{32.2} a_r$ (1)

$\overset{+}{\leftarrow} \Sigma F_\theta = ma_\theta;$ _____ $= \dfrac{2}{32.2} a_\theta$ (2)

At $\theta = \pi/2$

$\dot{\theta} =$ _____

$\ddot{\theta} =$ _____

$r =$ _____

$\dot{r} =$ _____

$\ddot{r} =$ _____

$a_r = \ddot{r} - r\dot{\theta}^2 = 0 - \pi(4)^2 = -50.27$ ft/s²

$a_\theta = r\ddot{\theta} + 2\dot{r}\dot{\theta} = 0 + 2(8)(4) = 64$ ft/s²

Substituting into Eqs. (1) and (2) and solving yields

$P = 1.66$ lb, $N_s = 4.78$ lb   Ans.

13-15. Rod OA rotates counterclockwise with a constant angular rate of $\dot{\theta} = 5$ rad/s. The double collar B is pin-connected together such that one collar slides over the rotating rod and the other slides over the *horizontal* curved rod, of which the shape is a limaçon described by the equation $r = 1.5(2 - \cos\theta)$ ft. If both collars weigh 0.75 lb, determine the normal force which the curved path exerts on one of the collars, and the force that OA exerts on the other collar at the instant $\theta = 90°$.

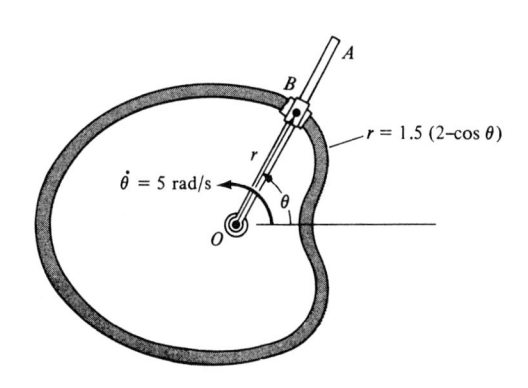

Solution

$$r = 1.5(2 - \cos\theta)$$

$$\frac{dr}{d\theta} = \underline{\hspace{3cm}}$$

at $\theta = \pi/2$

$\tan\psi = \underline{\hspace{3cm}}$

$\psi = 63.43°$

$+\uparrow \Sigma F_r = ma_r;$ _____ $= \dfrac{0.75}{32.2} a_r$

$\overset{+}{\leftarrow} \Sigma F_\theta = ma_\theta;$ _____ $= \dfrac{0.75}{32.2} a_\theta$

at $\theta = \pi/2$

$\dot\theta = 5$
$\ddot\theta = 0$
$r = 1.5(2 - \cos\theta) = 3$
$\dot r = 1.5(\sin\theta)\dot\theta = 7.5$
$\ddot r = 1.5(\cos\theta)\dot\theta^2 + 1.5\sin\theta(\ddot\theta) = 0$
$a_r = \ddot r - r(\dot\theta)^2 = 0 - 3(5)^2 = -75$
$a_\theta = r\ddot\theta + 2\dot r\dot\theta = 0 + 2(7.5)(5) = 75$

Solving;

$$N_C = 1.95 \text{ lb}, \quad F = 0.873 \text{ lb} \qquad \text{Ans.}$$

13-16. A truck T has a weight of 8,000 lb and is traveling along a portion of a road defined by the lemniscate $r^2 = 0.2(10^6) \cos 2\theta$, where r is measured in meters and θ is in radians. If the truck maintains a constant speed of $v_T = 4$ ft/s, determine the magnitude of the resultant frictional force which must be exerted by all the wheels to maintain the motion when $\theta = 0$.

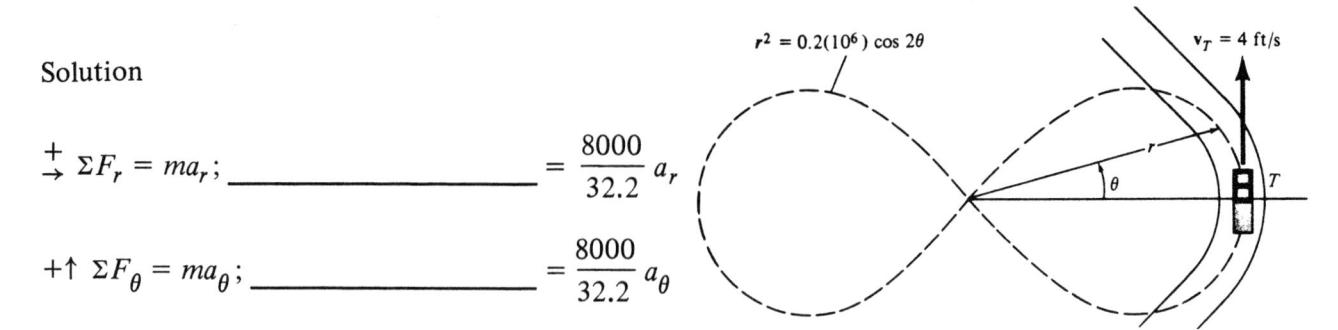

Solution

$\stackrel{+}{\rightarrow} \Sigma F_r = ma_r;$ _____ $= \dfrac{8000}{32.2} a_r$

$+\uparrow \Sigma F_\theta = ma_\theta;$ _____ $= \dfrac{8000}{32.2} a_\theta$

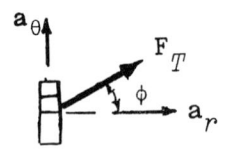

Kinematics

$r^2 = (0.2)(10^6) \cos 2\theta$
$(2r)\dot{r} = (-2)(0.2(10^6)) \sin 2\theta \, \dot{\theta}$
$(2\dot{r})\dot{r} + (2r)\ddot{r} = [(-4)(0.2(10^6)) \cos 2\theta \, \dot{\theta}]\dot{\theta} - 2(0.2(10^6)) \sin 2\theta \, \ddot{\theta}$

When $\theta = 0$

$\qquad r = 447.21$ (1)
$\qquad \dot{r} = 0$ (2)
$\qquad 2(447.21)\ddot{r} = (-4)(0.2(10^6))\dot{\theta}^2$ (3)

Since

$$v_T^2 = \dot{r}^2 + (r\dot{\theta})^2$$

$$(4)^2 = \dot{r}^2 + (r\dot{\theta})^2$$

Differentiating, we obtain

$$0 = 2\dot{r}\ddot{r} + 2r\dot{r}\dot{\theta}^2 + 2r^2\dot{\theta}\ddot{\theta}$$

Using the results of Eq. (1) and (2)

$$\dot{\theta} = 0.0089 \text{ rad/s} \quad \text{and} \quad \ddot{\theta} = 0$$

Therefore Eq. (3) becomes

$$\ddot{r} = -0.071$$

The acceleration components are therefore

$\qquad a_r = \ddot{r} - r\dot{\theta}^2$
$\qquad a_r = -0.071 - (447.21)(0.0089)^2 = -0.106 \text{ ft/s}^2$
$\qquad a_\theta = r\ddot{\theta} + 2\dot{r}\dot{\theta}$
$\qquad a_\theta = 0$

Thus

$$F_T = \left(\dfrac{8000}{32.2}\right)(-0.106) = -26.3 \text{ lb} \qquad \textit{Ans.}$$

$\qquad \phi = 0$

14 Kinetics of a Particle: Work and Energy

Principle of Work and Energy

14-1. A car having a mass of 2 Mg strikes a smooth, rigid sign post with an initial speed of 30 km/h. To stop the car, the front end horizontally deforms 0.2 m. If the car is free to roll during the collision, determine the *average* horizontal collision force causing the deformation.

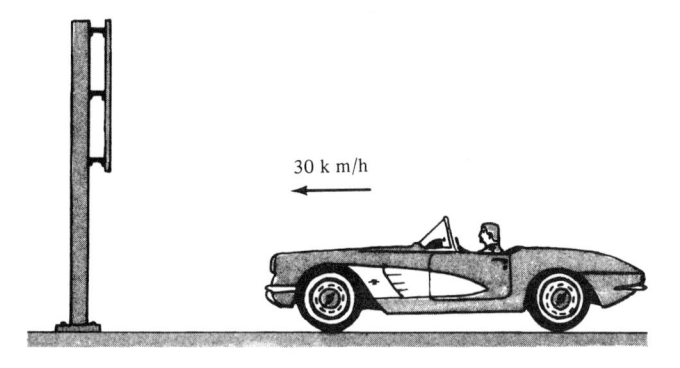

Solution

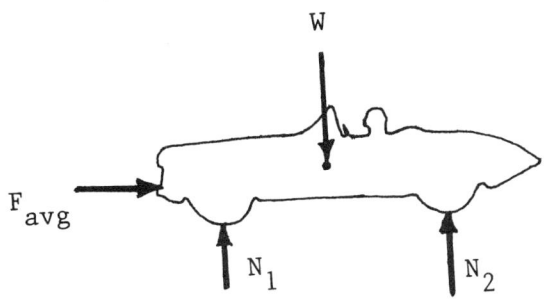

$v_1 = 30$ km/h $= 8.33$ m/s

$\Delta s = 0.20$ m

Apply the equation of work and energy

$$T_1 + \Sigma U_{1-2} = T_2$$

$F_{avg} = 347$ kN $\qquad$ *Ans.*

14-2. A car is equipped with a bumper B designed to absorb collisions. The bumper is mounted to the car using pieces of flexible tubing T. Upon collision with a rigid barrier A, a constant horizontal force **F** is developed which causes a car deceleration of $3g = 29.43$ m/s² (the highest safe deceleration for a passenger without a seatbelt). If the car and passenger have a total mass of 1.5 Mg and the car is initially coasting with a speed of 1.5 m/s, compute the magnitude of **F** needed to stop the car and the deformation x of the bumper tubing.

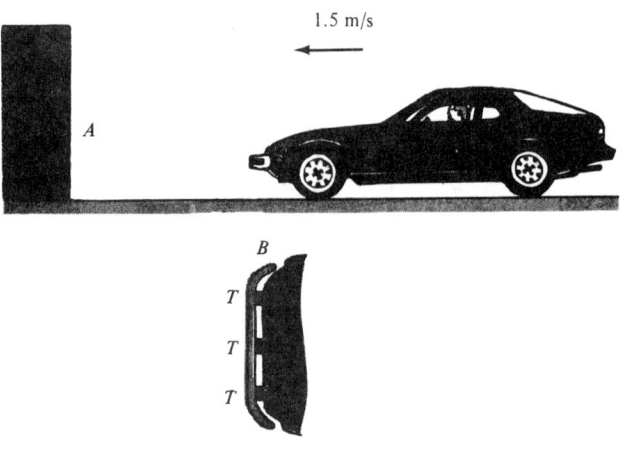

Solution

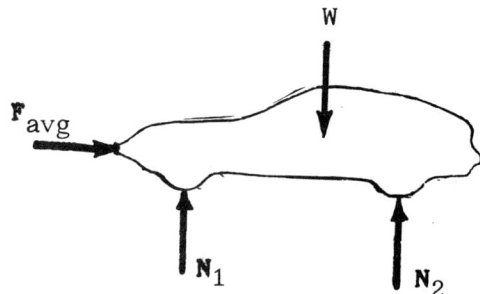

The average force needed to decelerate the car is

$$\overset{+}{\rightarrow} \Sigma F_x = ma_x;$$ _____

$F_{avg} = 44\ 145 = 44.1$ kN   Ans.

The deformation is determined from the equation of work and energy.

$$T_1 + \Sigma U_{1-2} = T_2$$

_____

$x = 0.0382 = 38.2$ mm   Ans.

14-3. A car, assumed to be rigid and having a mass of 800 kg, strikes a barrel-barrier installation without the driver applying the brakes. From experiments, the magnitude of the force of resistance F_r, created by deforming the barrels successively, is shown as a function of vehicle penetration. If the car strikes the barrier traveling at $v_c = 70$ km/h, determine approximately the distance s to which the car penetrates the barrier.

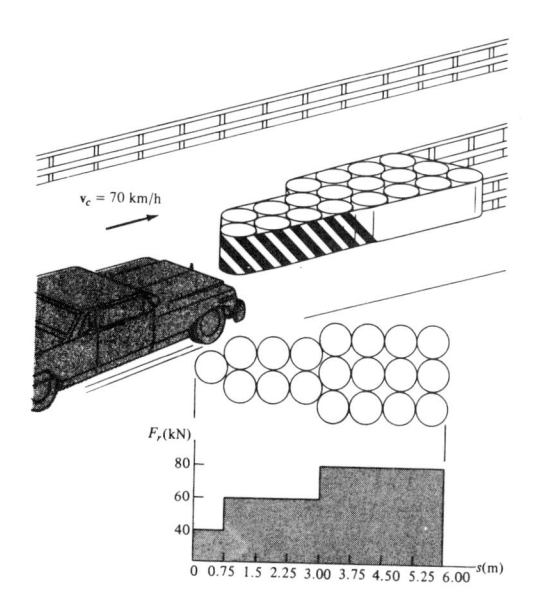

Solution

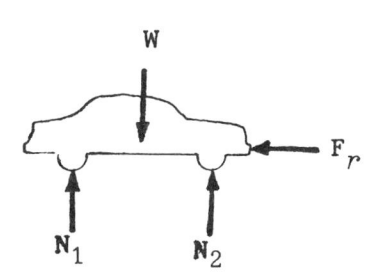

$$v_1 = 70 \text{ km/h} = 19.44 \text{ m/s}$$

Negative work is done by F_r as the car penetrates the barrier. This work is equal to the area under the $F_r - s$ graph.

$$\Sigma U_{1-2} = T_2 - T_1$$

Thus

$$U_{1-2} = \frac{1}{2}(800)(19.44)^2$$

$$U_{1-2} = 151.2 \text{ kJ}$$

From the graph, this work in terms of s is equal to an area

$$151.2 = (0.75)(40) + \underline{}$$

$$s = 2.77 \text{ m} \qquad \qquad Ans.$$

14-4. When at A the bicyclist has a speed of $v_A = 4$ ft/s. If he coasts without pedaling from the top of the hill at A to the shore of B and then leaps off the shore, determine his speed at B and the distance x where he strikes the water at C. The rider and his bicycle have a total weight of 150 lb. Neglect the size of the bicycle and wind resistance.

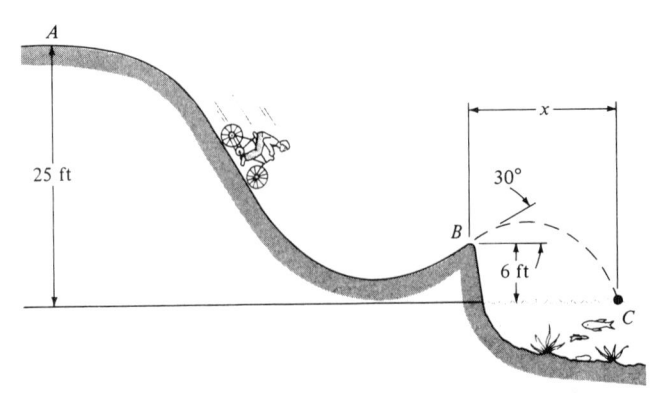

Solution

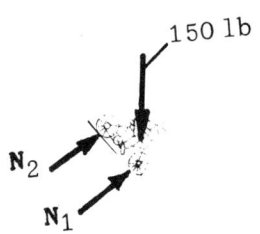

Determine the launch speed at B

$$T_A + \Sigma U_{AB} = T_B$$

$$v_B = 35.2 \text{ ft/s} \qquad \textit{Ans.}$$

$$(v_B)_x = 35.2 \cos 30° = 30.49$$

$$(v_B)_y = 35.2 \sin 30° = 17.60$$

$$\overset{+}{\rightarrow} \; x = 30.49t$$

$$+\uparrow \; s_C = s_B + (v_B)_y t + \frac{1}{2} a_c t^2$$

$$-6 = 0 + 17.60t + \frac{1}{2}(-32.2)t^2$$

Solving for the positive root:

$$t = 1.37 \text{ s}$$

Thus

$$x = (30.49)(1.37) = 41.7 \text{ ft} \qquad \textit{Ans.}$$

14-5. The coefficient of friction between the 2-lb block and the surface is $\mu = 0.2$. The block is acted upon by a horizontal force of $P = 15$ lb and has a speed of 6 ft/s when it is at point A. Determine the maximum deformation of the outer spring B at the instant the block comes to rest. Spring B has a stiffness of $k_B = 20$ lb/ft and the "nested" spring C has a stiffness of $k_C = 40$ lb/ft.

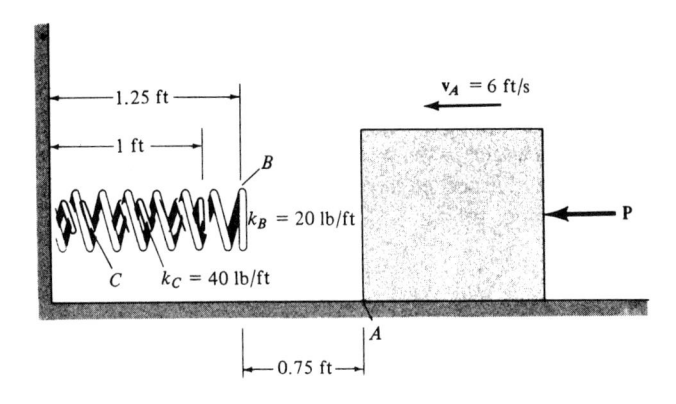

Solution

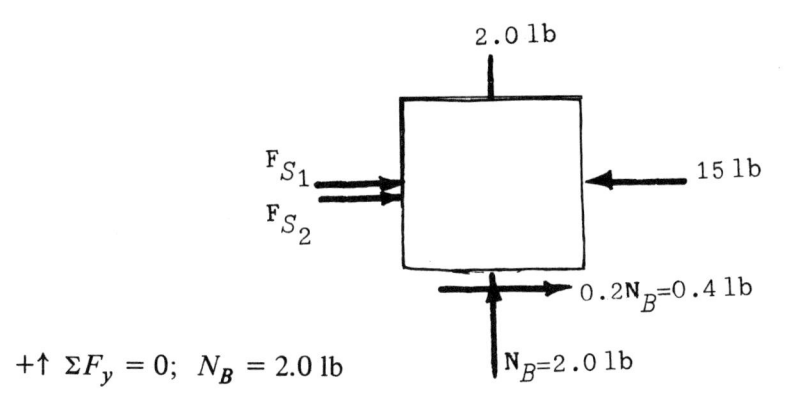

$+\uparrow \Sigma F_y = 0; \quad N_B = 2.0$ lb

Assume both springs are compressed.

$$T_1 + \Sigma U_{1-2} = T_2$$

$$-30x^2 + 24.6x + 10.82 = 0$$

Solving:

$x = 1.14$ ft (compression of B)   *Ans.*

$x - 0.25 = 0.89$ ft (compression of C)

14-6. The "flying car" is a ride at an amusement park, which consists of a car having wheels that roll along a track mounted on a rotating drum. Motion of the car is created by applying the car's brake, thereby gripping the car to the track and allowing it to move with a speed of $v_t = 3$ m/s. If the rider applies the brake when going from B to A and then releases it at the top of the drum, A, so that the car coasts freely down along the track to B ($\theta = \pi$ rad), determine the speed of the car at B and the normal reaction which the drum exerts on the car at B. The rider and car have a total mass of $m = 250$ kg and the center of mass of the car and rider moves along a circular path of radius $r = 8$ m.

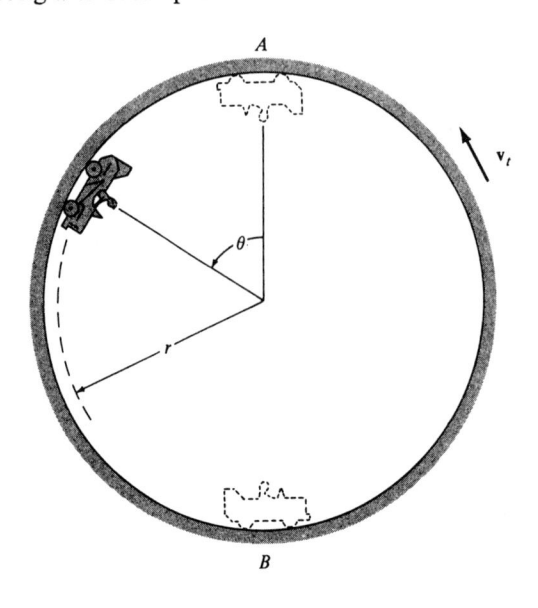

Solution

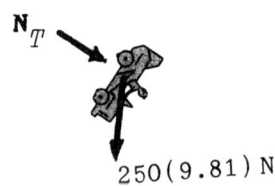

$$T_A + \Sigma U_{A\text{-}B} = T_B$$

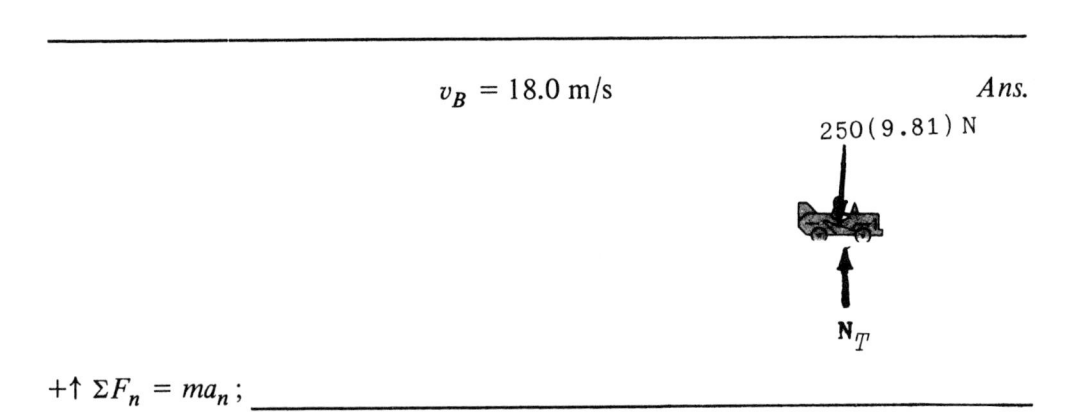

$$v_B = 18.0 \text{ m/s} \quad Ans.$$

$$+\uparrow \Sigma F_n = ma_n;$$

$$N_T = 12.5 \text{ kN} \quad Ans.$$

Power and Efficiency

14-7. A motor hoists a 50-kg crate at constant speed to a height of $h = 6$ m in 3 s. If the indicated power of the motor is 4 kW, determine the motor's efficiency.

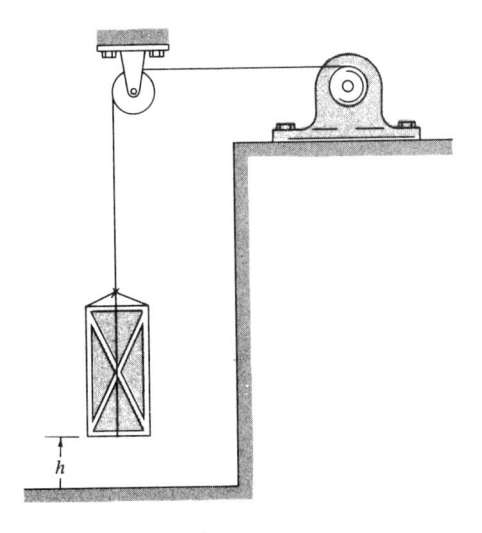

Solution

Work is done against gravity

$U_{1-2} =$ _____

$U_{1-2} = 2943$ J

$P_{out} =$ _____

$P_{out} = 981$ W

$\epsilon = \dfrac{P_{out}}{P_{in}} =$ _____

$\epsilon = 0.245$ *Ans.*

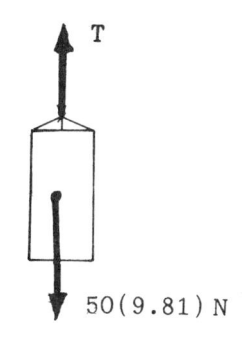

14-8. A truck has a weight of 25,000 lb and an engine which transmits a power of 350 hp to *all* the wheels. Assuming that the wheels do not slip on the ground, determine the angle θ of the largest incline the truck can climb at a constant speed of $v = 50$ ft/s.

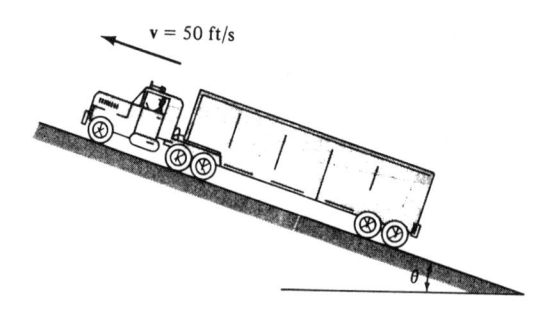

Solution

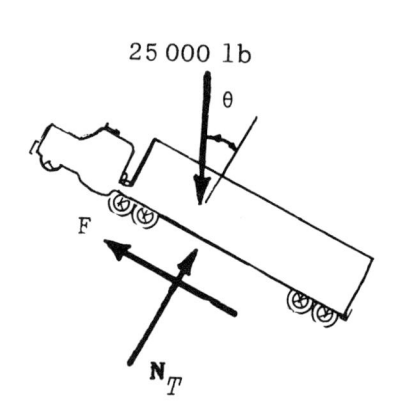

$\nwarrow+ \Sigma F_x = 0; \quad F = 25\,000 \sin\theta$ lb

$\mathbf{P} = \mathbf{F} \cdot \mathbf{v}$

$\theta = 8.86°$ Ans.

14-9. The elevator E and its freight have a total mass of 400 kg. Hoisting is provided by the motor M and the 60-kg block C. If the motor has an efficiency of $\epsilon = 0.6$, determine the power that must be supplied to the motor when the elevator is hoisted upward at a constant speed of $v_E = 4$ m/s.

Solution

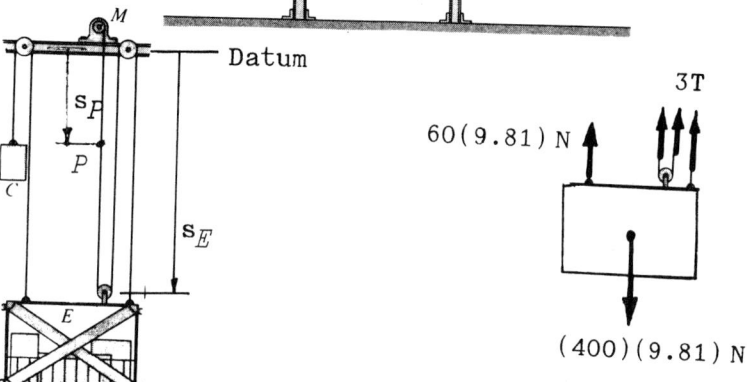

Since $a = 0$,

$$+\uparrow \Sigma F_y = 0$$

$$T = 1111.8 \text{ N}$$

The position equation is

Taking the time derivative,

$$3v_E = v_P$$

Since

$$v_E = 4 \text{ m/s} \uparrow$$
$$v_P = 12 \text{ m/s} \uparrow$$

Thus

$$P_2 = \frac{\mathbf{T} \cdot \mathbf{v}_P}{\epsilon} = \frac{1111.8(12)}{0.6}$$

$$P_2 = 22.2 \text{ kW} \hspace{2cm} Ans.$$

14-10. An electric train car, having a mass of 25 Mg, travels up a 10° incline with a constant speed of 80 km/h. Determine the power required to overcome the force of gravity.

Solution

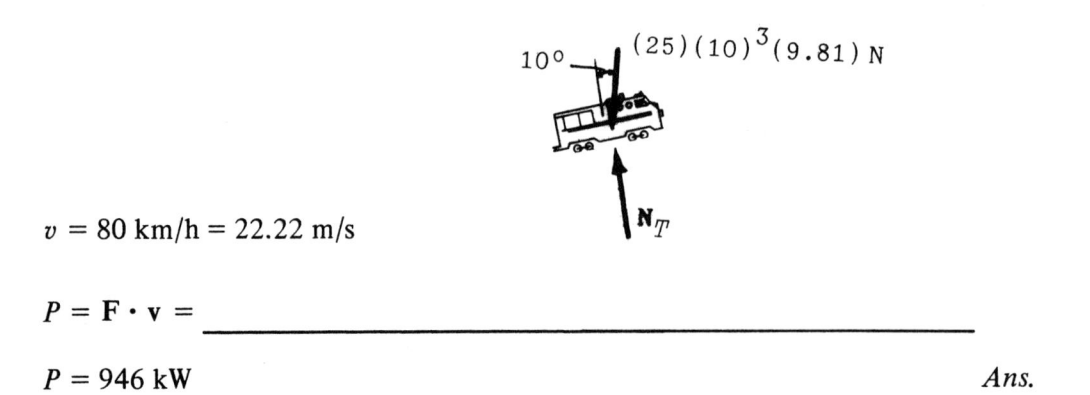

$v = 80$ km/h $= 22.22$ m/s

$P = \mathbf{F} \cdot \mathbf{v} = $ _____

$P = 946$ kW   Ans.

Conservation of Energy Theorem

14-11. The block has a weight of 1.5 lb and slides along the smooth chute AB. It is released from rest at A, which has coordinates of $A(5 \text{ ft}, 0, 10 \text{ ft})$. Determine the speed at which it slides off at B, which has coordinates of $B(0, 8 \text{ ft}, 0)$.

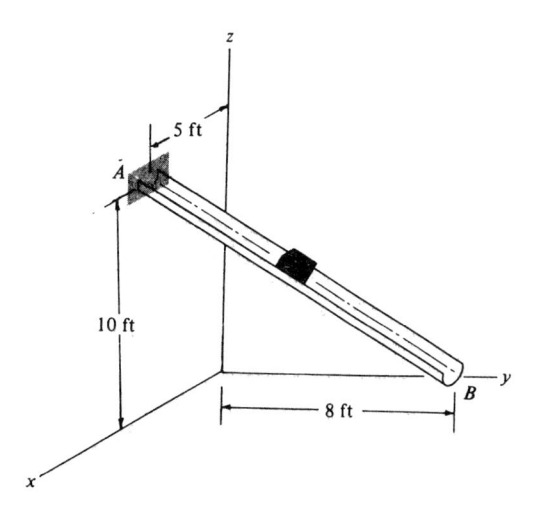

Solution

Place the datum at B.

$$T_A + V_A = T_B + V_B$$

$$v_B = 25.4 \text{ ft/s} \qquad Ans.$$

14-12. The firing mechanism of a pinball machine consists of a plunger P having a mass of 0.25 kg and a spring of stiffness $k = 300$ N/m. When $s = 0$, the spring is compressed 50 mm. If the arm is pulled back such that $s = 100$ mm and released, determine the speed of the 0.3 kg pinball B *just before* the plunger strikes the stop, i.e., $s = 0$. Assume all surfaces of contact to be smooth. The ball moves in the horizontal plane. Neglect friction and the rolling motion of the ball.

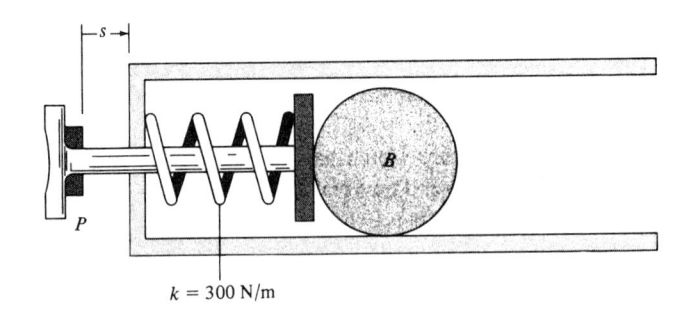

Solution

$$T_1 + V_1 = T_2 + V_2$$

$$v_2 = 3.30 \text{ m/s} \qquad Ans.$$

14-13. The block A having a weight of 1.5 lb slides in the smooth horizontal slot. If the block is drawn back so that $s = 1.5$ ft and released from rest, determine its speed at the instant $s = 0$. Each of the two springs has a stiffness of $k = 150$ lb/ft and an unstretched length of 0.5 ft.

Solution

$x_1 = (\sqrt{(2)^2 + (1.5)^2} - 0.5)$ ft

$x_2 = (2 - 0.5)$ ft

$T_1 + V_1 = T_2 + V_2$

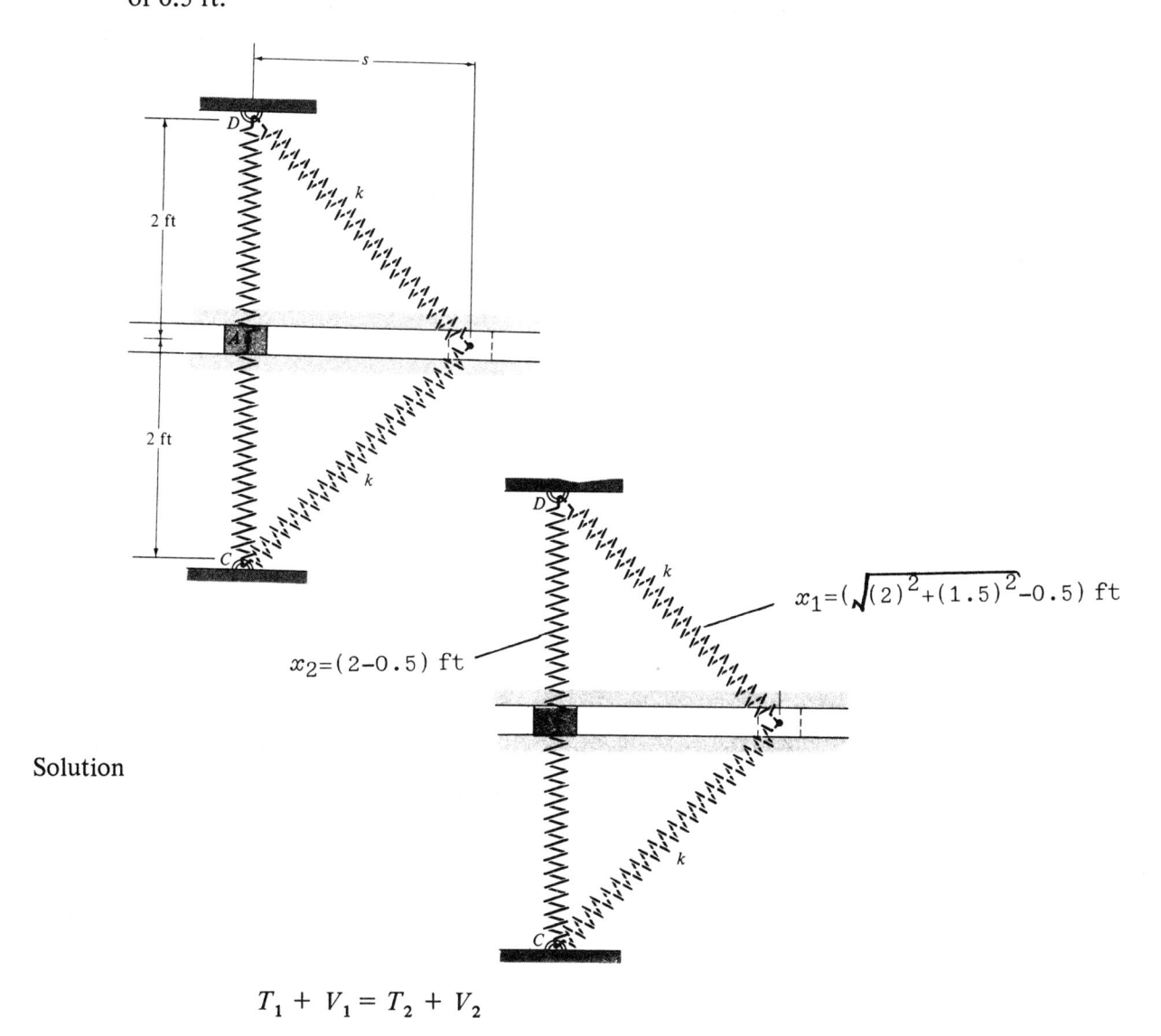

$v_2 = 106$ ft/s   Ans.

14-14. The roller-coaster car has a speed of 15 ft/s when it is at the crest of a vertical parabolic track. Compute the velocity and the normal force it exerts on the track when it reaches point B. Neglect friction and the mass of the wheels. The total weight of the car and the passengers is 350 lb.

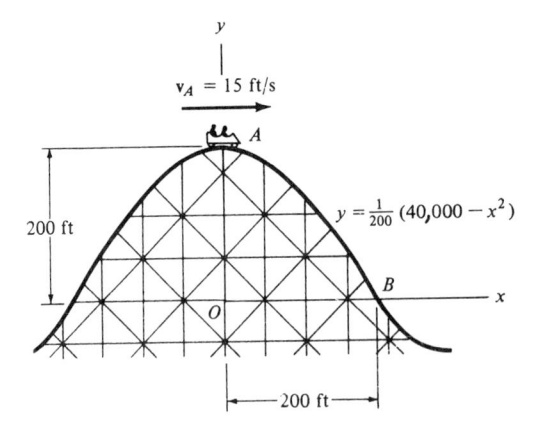

Solution

Establish the datum at A

$$T_A + V_A = T_B + V_B$$

$$v_B = 114 \text{ ft/s} \quad \text{Ans.}$$

$$y = \frac{1}{200}(40{,}000 - x^2)$$

$$\frac{dy}{dx} = -\frac{1}{100}x$$

At $x = 200$

$$\frac{dy}{dx} = \tan\theta = -2$$

$$\theta = -63.43°$$

$$\frac{1}{\rho} = \left| \frac{\frac{d^2y}{dx^2}}{\left[1 + \left(\frac{dy}{dx}\right)^2\right]^{3/2}} \right|$$

$$= \left| \frac{-\frac{1}{100}}{[1 + (-2)^2]^{3/2}} \right|$$

$$\rho = 1118.0 \text{ ft}$$

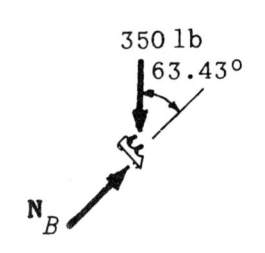

$$+\nwarrow \Sigma F_n = ma_n,$$

$$N_B = 29.1 \text{ lb} \quad \text{Ans.}$$

14-15. The car C and its contents have a weight of 600 lb, whereas block B has a weight of 200 lb. If the car is released from rest, determine its speed when it travels 30 ft down the 20° incline.

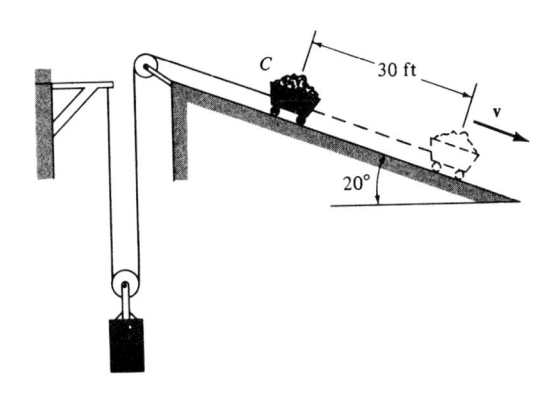

Solution

Kinematics: $\qquad 2s_B + s_C = l$

For a displacement

$$2|\Delta s_B| = |\Delta s_C|$$

also

$$2|v_B| = |v_C|$$

When $\Delta s_C = 30$ ft, $\Delta s_B = 15$ ft

Establishing two datums at the initial elevations of the car and block, respectively, we have

$$T_1 + V_1 = T_2 + V_2$$

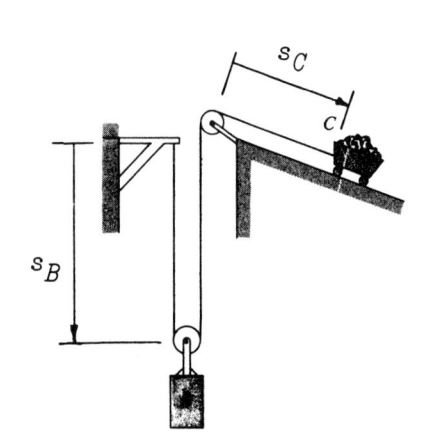

$$v_C = 17.7 \text{ ft/s} \qquad Ans.$$

15 Kinetics of a Particle: Impulse and Momentum

Principle of Linear Impulse and Momentum

15-1. A hockey puck is traveling to the left with a velocity of $v_1 = 10$ m/s when it is struck by a hockey stick and given a velocity of $v_2 = 20$ m/s as shown. Determine the magnitude of the net impulse exerted by the hockey stick on the puck. The puck has a mass of 0.2 kg.

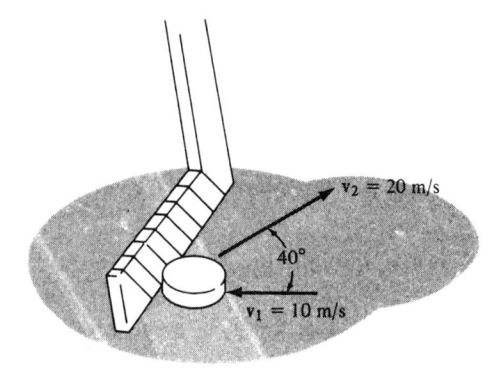

Solution

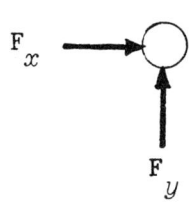

$\xrightarrow{+}\ m(v_x)_1 + \Sigma \int F_x\,dt = m(v_x)_2$

$\int F_x\,dt = 5.06$ N · s

$+\uparrow\ m(v_y)_1 + \Sigma \int F_y\,dt = m(v_y)_2$

$\int F_y\,dt = 2.57$ N · s

$\text{Imp} = \sqrt{(5.06)^2 + (2.57)^2} = 5.68$ N · s   *Ans.*

15-2. A golf ball having a mass of 40 g is struck such that it has an initial velocity of 200 m/s as shown. Determine the horizontal and vertical components of the impulse given to the ball.

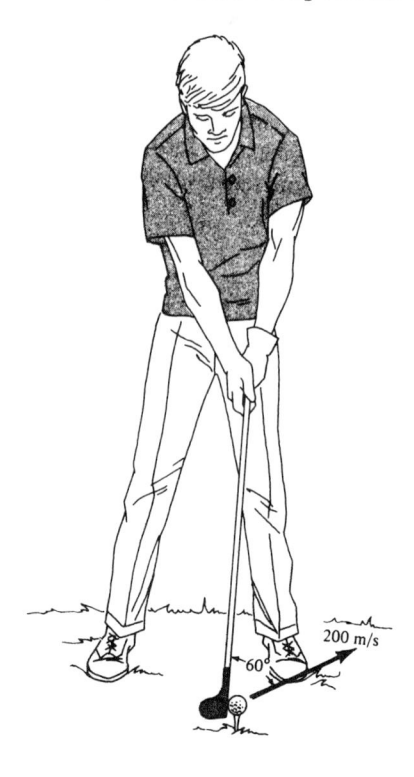

Solution

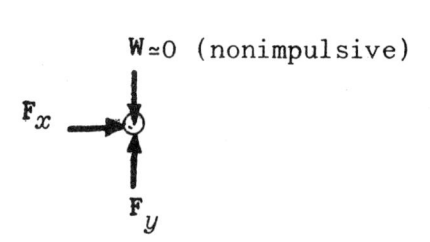

$W \approx 0$ (nonimpulsive)

$$\xrightarrow{+} \quad m(v_x)_1 + \Sigma \int F_x \, dt = m(v_x)_2$$

$\int F_x \, dt = 6.93 \text{ N} \cdot \text{s}$ Ans.

$+\uparrow \quad m(v_y)_1 + \Sigma \int F_y \, dt = m(v_y)_2$

$\int F_y \, dt = 4.00 \text{ N} \cdot \text{s}$ Ans.

15-3. Determine the velocities of blocks A and B 2 s after they are released from rest. Neglect the mass of the pulleys and cables.

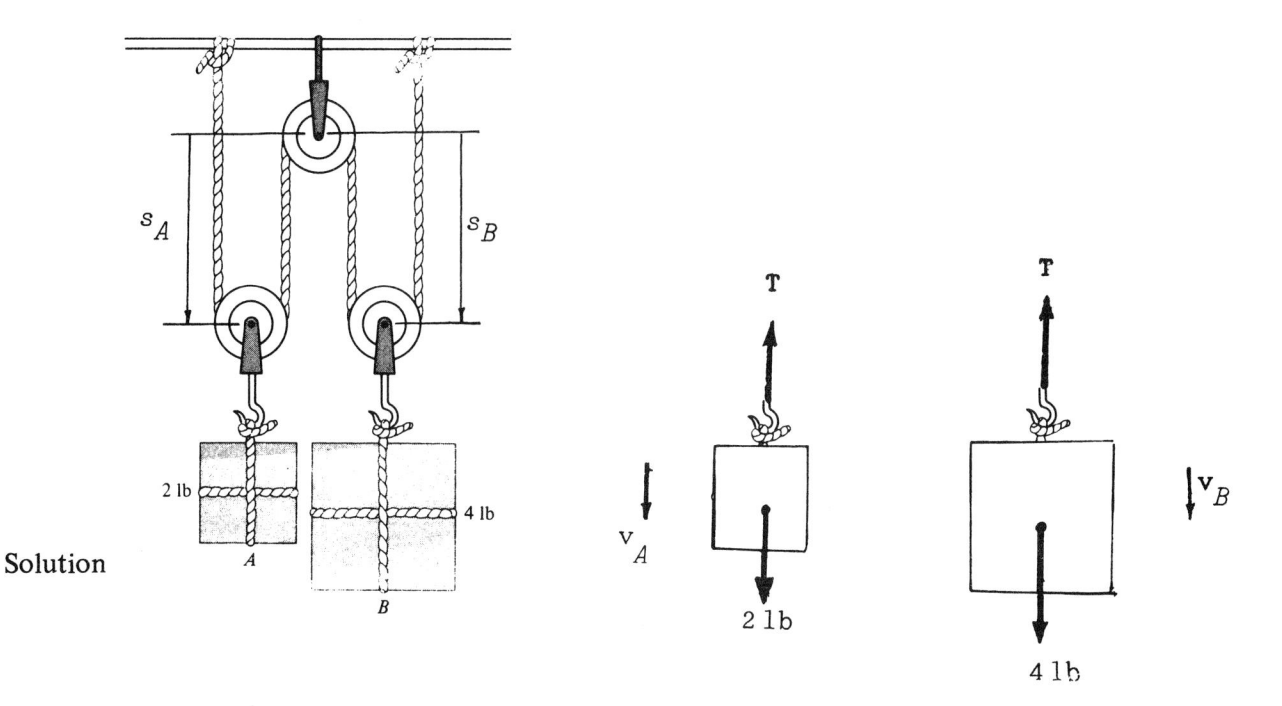

Solution

Block A:

$$+\downarrow \quad m(v_A)_1 + \Sigma \int F_y \, dt = m(v_A)_2$$

Block B:

$$+\downarrow \quad m(v_B)_1 + \Sigma \int F_y \, dt = m(v_B)_2$$

Position Equation

$$v_A = -v_B \tag{3}$$

Note that v_A and v_B are *both* assumed to be directed downward in *all* the above equations. Solving:

$$T = 2.61 \text{ lb}$$

$$v_A = 21.5 \text{ ft/s}\uparrow, \quad v_B = 21.5 \text{ ft/s}\downarrow \qquad \qquad Ans.$$

15-4. In cases of emergency, the gas actuator can be used to move a 75-kg block B by exploding a charge C near a pressurized cylinder of negligible mass. As a result of the explosion, the cylinder fractures and the released gas forces the front part of the cylinder, A, to move B forward, giving it a speed of 200 mm/s in 0.4 s. If the coefficient of friction between B and the floor is $\mu = 0.5$, determine the impulse that the actuator must impart to B.

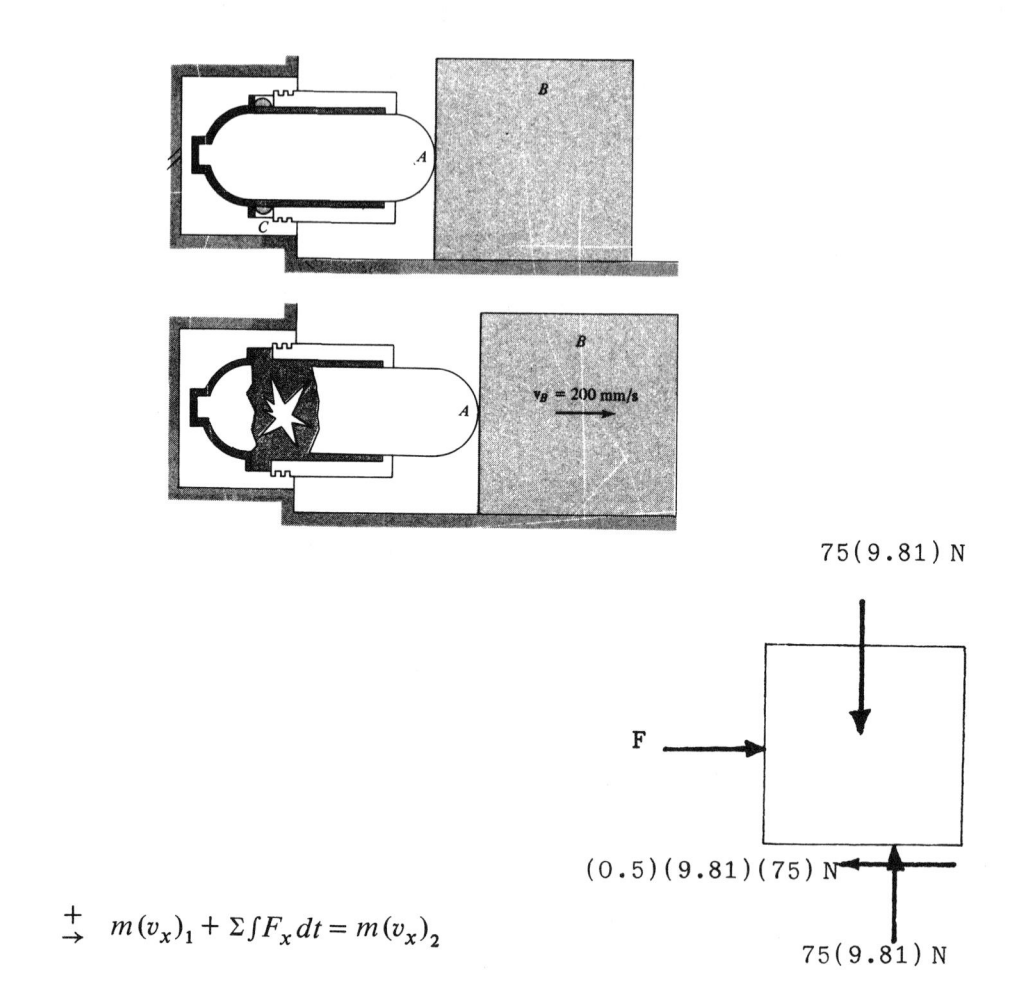

Solution

$\xrightarrow{+} \quad m(v_x)_1 + \Sigma \int F_x \, dt = m(v_x)_2$

$\int F \, dt = 162 \text{ N} \cdot \text{s}$ *Ans.*

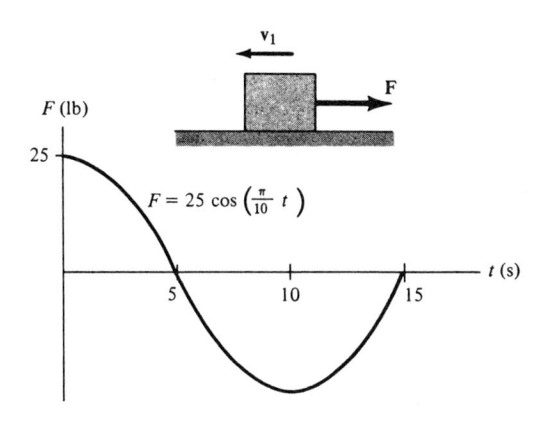

15-5. A 30-lb block is initially moving along a smooth horizontal surface with a speed of $v_1 = 6$ ft/s to the left. If it is acted upon by a force F, which varies in the manner shown, determine the velocity of the block in 15 s. The argument for the cosine is in radians.

$F = 25 \cos\left(\frac{\pi}{10} t\right)$

Solution

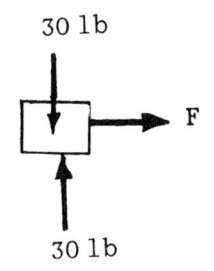

$$\overset{+}{\rightarrow} \quad m(v_x)_1 + \Sigma \int F_x \, dt = m(v_x)_2$$

$(v_x)_2 = 91.4$ ft/s $\leftarrow$ Ans.

15-6. The motor M pulls on the cable with a force **F** that has a magnitude which varies as shown on the graph. If the 15-kg crate is originally resting on the floor such that the cable tension is zero when the motor is turned on, determine the speed of the crate when $t = 6$ s. *Hint:* First determine the time needed to begin lifting the crate.

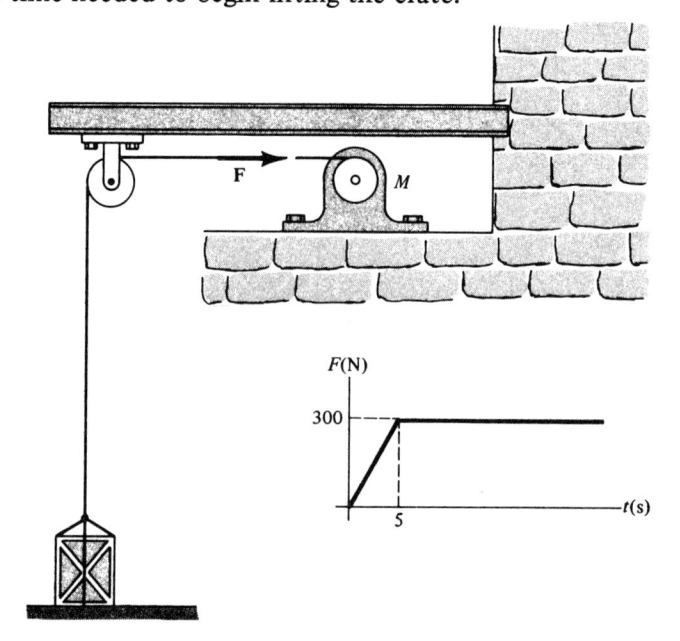

Solution

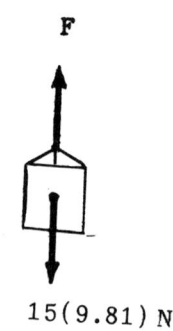

To lift the crate $F = 15(9.81) = 147.2$ N.

From the graph,

$$0 \leq t \leq 5 \text{ s}, \quad F = \frac{300}{5} t = 60t$$

Hence, the time to lift the crate is

$$t = \frac{147.2}{60} = 2.45 \text{ s}$$

The impulse of the motor is calculated as the area under the graph.

$$+\uparrow \quad m(v_y)_1 + \Sigma \int F_y \, dt = m(v_y)_2$$

$$(v_y)_2 = 23.2 \text{ m/s} \qquad \qquad Ans.$$

Conservation of Linear Momentum for a System of Particles

15-7. A rifle has a mass of 2.5 kg. If it is loosely gripped and a 1.5-g bullet is fired from it with a muzzle velocity of 1400 m/s, determine the recoil velocity of the rifle just after firing.

Solution

$$\xrightarrow{+} \Sigma mv_1 = \Sigma mv_2$$

$(v_R)_2 = 0.840$ m/s   Ans.

15-8. A 0.6-kg brick is thrown into a 25-kg wagon which is initially at rest. If, upon entering, the brick has a velocity of 10 m/s as shown, determine the final velocity of the wagon.

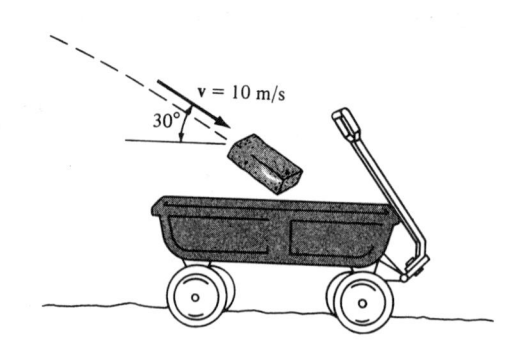

Solution

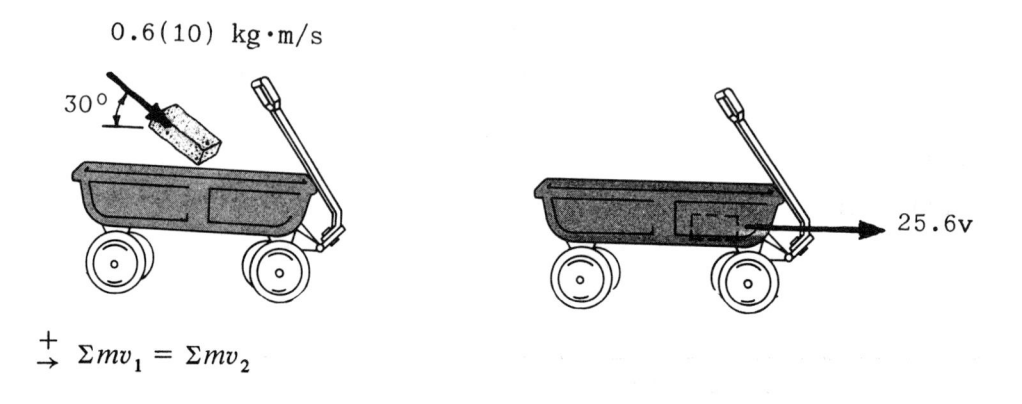

$\xrightarrow{+} \Sigma mv_1 = \Sigma mv_2$

$v = 0.203$ m/s   Ans.

15-9. A girl having a weight of 40 lb slides down the smooth slide onto the surface of a 20-lb wagon. Determine the speed of the wagon at the instant the girl stops sliding on it. If someone ties the wagon to the slide at B, determine the horizontal impulse the girl will exert at C in order to stop her motion. Neglect friction and assume that the girl starts from rest at the top of the slide, A.

Solution

Put datum at B

$$T_A + V_A = T_B + V_B$$

$$v_B = 31.08 \text{ ft/s}$$

girl and wagon

$$\overset{+}{\leftarrow} \Sigma m v_1 = \Sigma m v_2$$

$$v_2 = 20.7 \text{ ft/s} \qquad \qquad Ans.$$

girl

$$\overset{+}{\leftarrow} m v_1 + \Sigma \int F dt = m v_2$$

$$\int F dt = 38.6 \text{ lb} \cdot \text{s} \qquad \qquad Ans.$$

15-10. A boy, having a weight of 90 lb, jumps off a wagon with a relative velocity of $v_{b/w} = 6$ ft/s. If the angle of jump is 30°, determine the horizontal velocity $(v_w)_2$ of the wagon just after the jump. Originally both the wagon and the boy are at rest. Also, compute the total average impulsive force that all four wheels of the wagon exert on the ground if the boy jumps off in $\Delta t = 0.8$ s. The wagon has a weight of 20 lb.

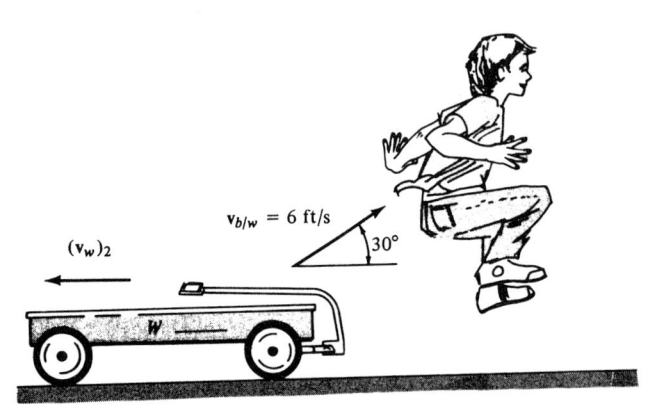

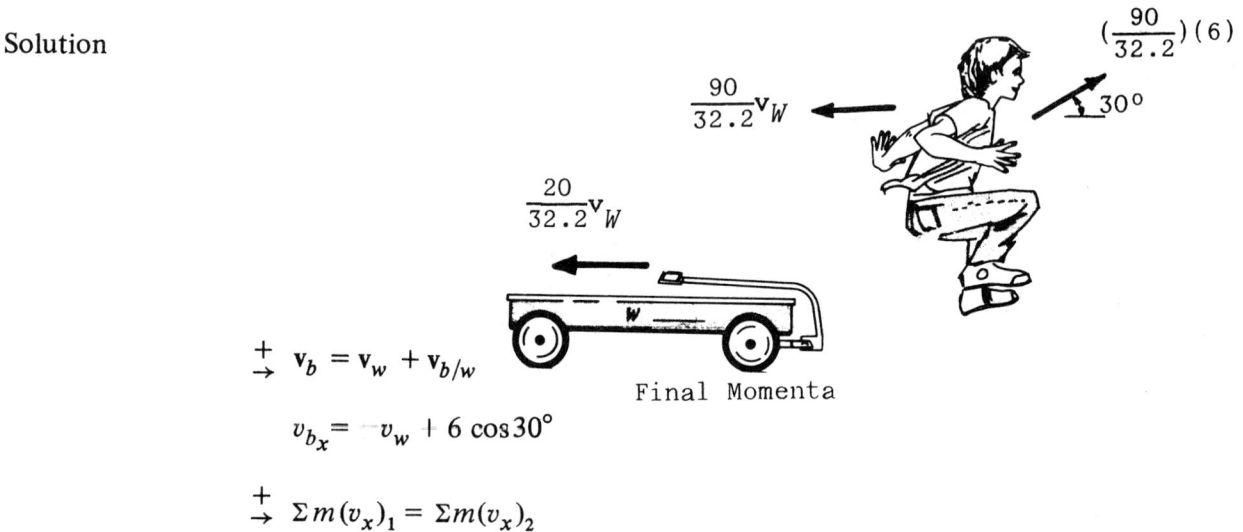

Solution

$\overset{+}{\rightarrow} \mathbf{v}_b = \mathbf{v}_w + \mathbf{v}_{b/w}$

$v_{b_x} = -v_w + 6\cos 30°$

$\overset{+}{\rightarrow} \Sigma m(v_x)_1 = \Sigma m(v_x)_2$

$v_w = 4.25$ ft/s *Ans.*

$+\uparrow \Sigma m(v_y)_1 + \Sigma \int F_y\, dt = \Sigma m(v_y)_2$

$N_w = 120$ lb *Ans.*

15-11. The two handcars A and B each have a mass of 80 kg. If the man C has a mass of 70 kg and jumps from A with a horizontal *relative* velocity of $v_{C/A} = 2$ m/s and lands on B, determine the velocity of each car after the jump. Neglect the effects of rolling resistance.

Solution

$\xrightarrow{+} \Sigma mv_1 = \Sigma mv_2$

$v_A = 0.933$ m/s   Ans.

$\xrightarrow{+} \Sigma mv_1 = \Sigma mv_2$

$v_B = 0.498$ m/s   Ans.

15-12. A man wearing ice skates throws an 8-kg block with an initial velocity of 2 m/s, measured relative to himself, in the direction shown. If he is originally at rest and completes the throw in 1.5 s while keeping his legs rigid, determine the horizontal velocity of the man just after releasing the block. What is the vertical reaction of both his skates on the ice during the throw? The man has a mass of 70 kg. Neglect friction and the motion of his arms.

Solution

Final Momenta

$\xrightarrow{+} \Sigma m v_1 = \Sigma m v_2$

$(v_m)_2 = 0.178$ m/s *Ans.*

$+\uparrow \Sigma m v_1 + \Sigma \int F dt = \Sigma m v_2$

$R_{Avg} = 770$ N *Ans.*

Impact

15-13. The drop hammer H has a weight of 900 lb and falls from rest $h = 3$ ft onto a forged anvil plate P that has a weight of 500 lb. The plate is mounted on a set of springs which have a combined stiffness of $k_T = 500$ lb/ft. Determine (a) the velocity of P and H just after collision and (b) the maximum compression in the springs caused by the impact. The coefficient of restitution between the hammer and the plate is $e = 0.6$. Neglect friction along the vertical guide posts A and B.

Solution

Just before impact put datum at initial position of hammer.

$$T_0 + V_0 = T_1 + V_1$$

$$(v_H)_1 = 13.90 \text{ ft/s}$$

Conservation of momentum will be applied since the force of the springs is nonimpulsive compared to the impact force.

$$+\downarrow \quad m_H(v_H)_1 + m_P(v_P)_1 = m_H(v_H)_2 + m_P(v_P)_2$$

$$13.9 = (v_H)_2 + 0.556(v_P)_2 \tag{1}$$

$$\downarrow + \quad e = \frac{(v_P)_2 - (v_H)_2}{(v_H)_1 - (v_P)_1}$$

$$8.34 = (v_P)_2 - (v_H)_2$$

Solving Eqs. (1) and (2);

$$(v_P)_2 = 14.29 \text{ ft/s}; \quad (v_H)_2 = 5.96 \text{ ft/s} \qquad \text{Ans.}$$

The initial compression in the springs is

$$F = kx; \quad 500 = 500(x_1)$$

$$x_1 = 1 \text{ ft}$$

Just after impact put datum at initial height of plate. Plate moves down x_2, spring compresses $x_2 + 1$.

$$T_1 + V_1 = T_2 + V_2$$

$$x_2 = 2.52 \text{ ft}$$

Total compression in spring is

$$x = x_2 + 1 = 3.52 \text{ ft} \qquad \text{Ans.}$$

15-14. Ball B has a mass of 0.75 kg and is moving forward with a velocity of $(v_B)_1 = 4$ m/s when it strikes the 2-kg block A, which is originally at rest. If the coefficient of restitution between the ball and the block is $e = 0.6$, compute (a) the velocity of A and B just after collision and (b) the distance block A slides before coming to rest. The coefficient of friction between the block and the surface is $\mu = 0.4$.

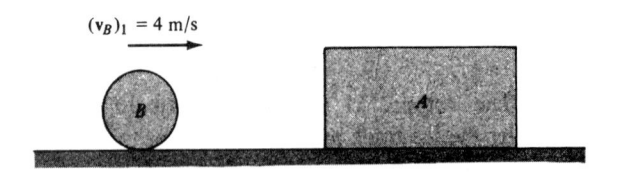

Solution

$\xrightarrow{+} \quad \Sigma m v_1 = \Sigma m v_2$

$$4 = -(v_B)_2 + 2.667(v_A)_2 \tag{1}$$

$\xrightarrow{+} \quad e = \dfrac{(v_A)_2 - (v_B)_2}{(v_B)_1 - (v_A)_1}$

$$(v_A)_2 + (v_B)_2 = 2.40 \tag{2}$$

Solving Eq. (1) and (2):

$(v_B)_2 = 0.655$ m/s; $(v_A)_2 = 1.75$ m/s **Ans.**

$(2)(9.81) = 19.62$ N

$(0.4)(19.62) = 7.848$ N

$N = 19.62$ N

$T_1 + \Sigma U_{1-2} = T_2$

$x = 0.390$ m **Ans.**

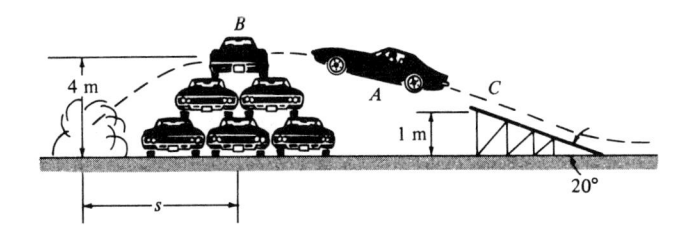

15-15. A stunt driver in car A travels in free flight off the edge of a ramp at C. At the point of maximum height he strikes car B. If the direct collision is perfectly plastic ($e = 0$), determine the required ramp speed v_C at the end of the ramp C, and the approximate distance s where both cars strike the ground. Each car has a mass of 3.5 Mg. Neglect the size of the cars in the calculation.

Solution

Determine the velocity of the car at C.

$$+\uparrow \quad (v_y)_2^2 = (v_y)_1^2 + 2a_c((s_y)_2 - (s_y)_1)$$

$$v_C = 22.43 \text{ m/s} = 80.8 \text{ km/h} \quad \text{Ans.}$$

The velocity just before collision is

$$(v_C)_x = 22.43 \cos 20° = 21.08 \text{ m/s}$$

The velocity of both cars just after collision is

$$\overset{+}{\leftarrow} \quad \Sigma m v_1 = \Sigma m v_2$$

$$v_2 = 10.54 \text{ m/s} \leftarrow$$

Time to fall is

$$+\downarrow \quad (s_y)_3 = (s_y)_2 + (v_y)_2 t + \frac{1}{2} a_c t^2$$

$$t = 0.903 \text{ s}$$

The distance s is

$$\overset{+}{\leftarrow} \quad (s_x)_3 = (v_x)_2 t$$

$$s = 9.52 \text{ m} \quad \text{Ans.}$$

15-16. Plates A and B each have a mass of 4 kg and are restricted to move along the frictionless guides. If the coefficient of restitution between the plates is $e = 0.7$, determine (a) the speed of both plates just after collision and (b) the maximum deflection of the spring. Plate A has a velocity of 4 m/s just before striking B. Plate B is originally at rest.

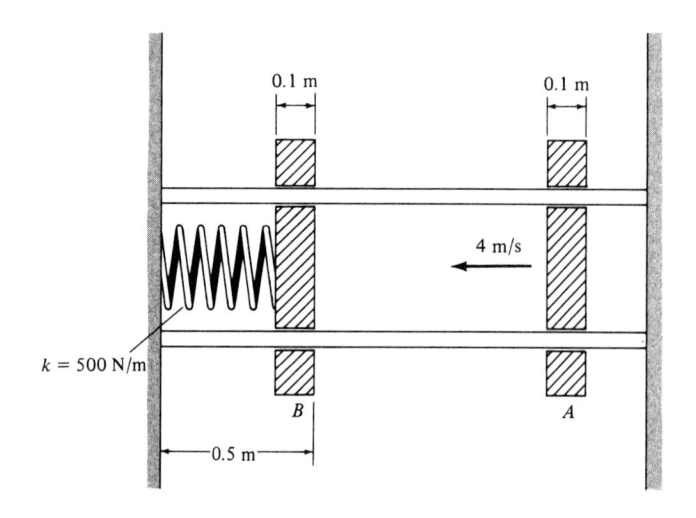

Solution

$$\overset{+}{\leftarrow} \; m_A(v_A)_1 + m_B(v_B)_1 = m_A(v_A)_2 + m_B(v_B)_2$$

(1)

$$\overset{+}{\leftarrow} \; e = \frac{(v_B)_2 - (v_A)_2}{(v_A)_1 - (v_B)_1}$$

(2)

Solving Eqs. (1) and (2),

$$(v_A)_2 = 0.600 \text{ m/s}, \quad (v_B)_2 = 3.40 \text{ m/s} \qquad \text{Ans.}$$

For plate B:

$$T_2 + V_2 = T_3 + V_3$$

$$x = 0.304 \text{ m} \qquad \text{Ans.}$$

15-17. Two coins A and B have the initial velocities shown just before they collide at point O. If they have weights of $W_A = 13.2(10^{-3})$ lb and $W_B = 6.6(10^{-3})$ lb and the surface upon which they slide is smooth, determine their speed just after impact. The coefficient of restitution is $e = 0.65$.

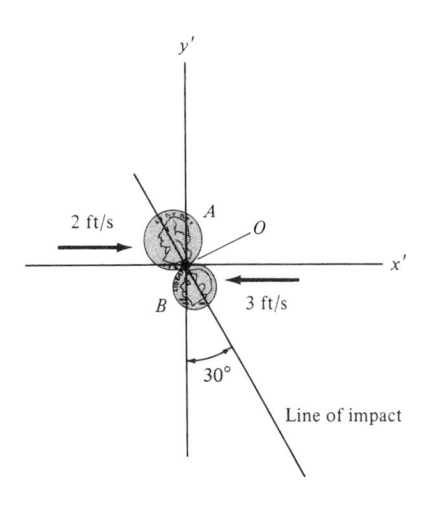

Solution

Along the line of impact (x axis)

$$+\downarrow \Sigma mv_1 = \Sigma mv_2$$

$$+\downarrow \quad e = \frac{(v_B)_{2x} - (v_A)_{2x}}{(v_A)_{1x} - (v_B)_{1x}}$$

Solving

$$(v_B)_{2x} = 1.25 \text{ ft/s}, \quad (v_A)_{2x} = -0.375 \text{ ft/s}$$

Coin A:

$$+\nearrow \Sigma m(v_A)_{1y} = \Sigma m(v_B)_{2y}$$

$$(v_A)_{2y} = 1.73 \text{ ft/s}$$

Coin B:

$$+\nearrow \Sigma m(v_B)_{1y} = \Sigma m(v_B)_{2y}$$

$$(v_B)_{2y} = 2.60 \text{ ft/s}$$

$$(v_B)_2 = \sqrt{(1.25)^2 + (2.60)^2} = 2.88 \text{ ft/s} \qquad Ans.$$

$$(v_A)_2 = \sqrt{(-0.375)^2 + (1.73)^2} = 1.77 \text{ ft/s} \qquad Ans.$$

84 Study Guide and Problems

Angular Momentum

15-18. The projectile having a mass of $m = 3$ kg is fired from a cannon with a muzzle velocity of $v_O = 500$ m/s. Determine the projectile's angular momentum about point O at the instant it is at the maximum height of its trajectory.

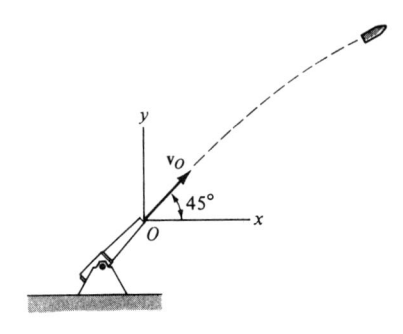

Solution

$$(v_O)_x = 500 \cos 45° = 353.5 \text{ m/s}$$

$$(v_O)_y = 500 \sin 45° = 353.5 \text{ m/s}$$

The maximum height of travel is

$$+\uparrow (v_y)^2 = (v_y)_0^2 + 2a_c(s_y - s_o)$$

_____

$h = 6369.1$

Since $(v_p)_x = (v_O)_x = 353.5$, then

$$H_O = r(mv)$$

$$H_O = \underline{\hspace{4cm}}$$

$$H_O = 6.75(10^6) \text{ kg} \cdot \text{m}^2/\text{s} \qquad \textit{Ans.}$$

15-19. A basket and its contents have a weight of 10 lb. Determine the speed at which the basket rises when $t = 3$ s, if initially a monkey having a weight of 20 lb begins to climb upward along the other end of the rope with a constant speed of $v_{m/r} = 2$ ft/s, measured relative to the rope. Neglect the mass of the pulley and rope.

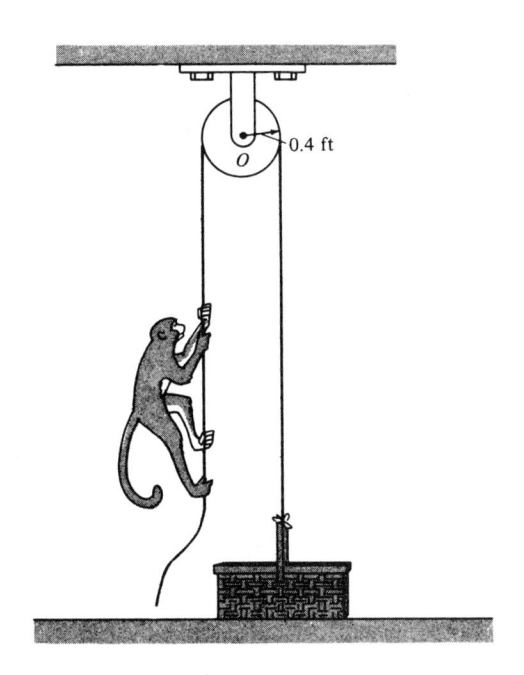

Solution

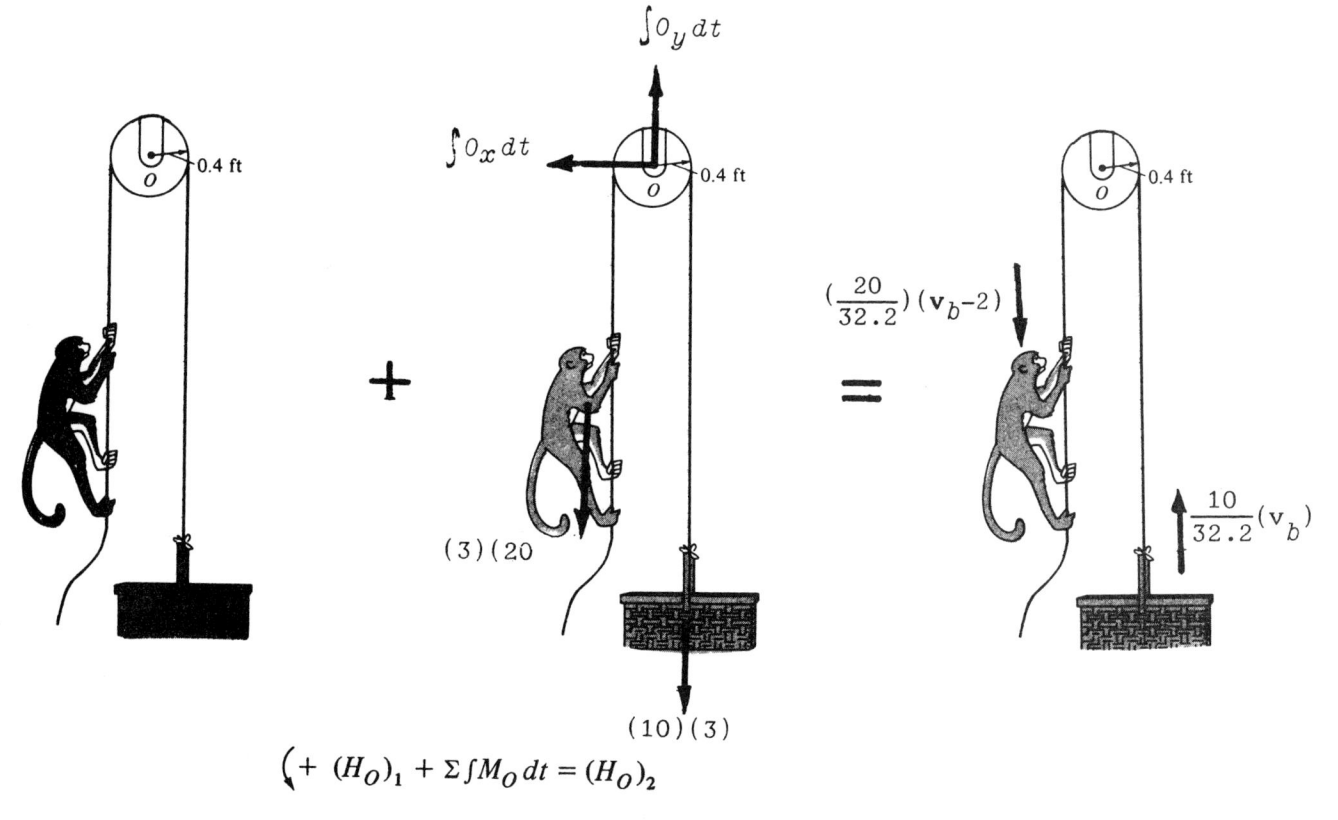

$$(+\circlearrowleft \; (H_O)_1 + \Sigma \int M_O \, dt = (H_O)_2$$

$v_b = 33.5$ ft/s Ans.

15-20. The two blocks A and B each have a mass of 500 g. The blocks are fixed to the horizontal rods and their initial velocity is 2 m/s in the direction shown. If a couple moment of $M = 0.8$ N·m is applied about CD of the frame, determine the speed of the block in 4 s. The mass of the supporting frame is negligible and it is free to rotate about CD.

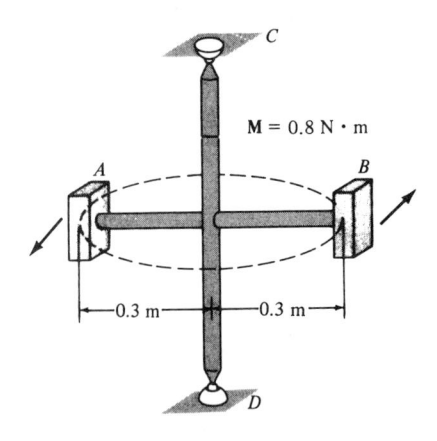

Solution

$$(H_{CD})_1 + \Sigma \int M_{CD}\, dt = (H_{CD})_2$$

$v_2 = 12.7$ m/s *Ans.*

15-21. A toboggan and rider, having a total mass of 150 kg, enter horizontally tangent to a 90° circular curve with a velocity of $v_A = 70$ km/h. If the track is flat and banked at an angle of 60°, determine the velocity v_A and the angle θ of "descent," measured from the horizontal in a vertical $x - z$ plane, at which the toboggan exits at B. Neglect friction in the calculation. The radius r_B equals 57 m.

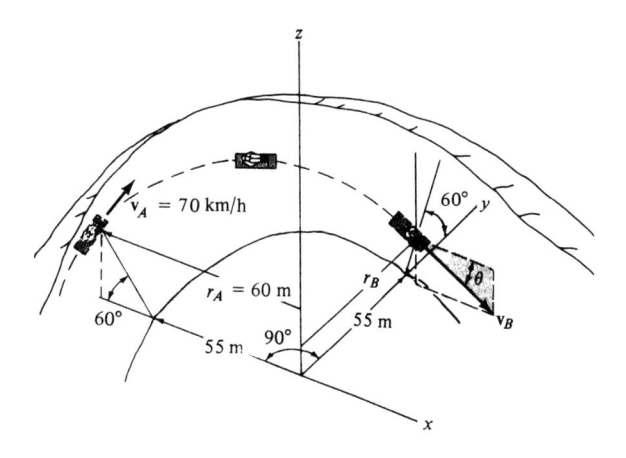

Solution

$$v_A = 70 \text{ km/h} = 19.44 \text{ m/s};$$

$$(H_A)_z = (H_B)_z$$

_____ (1)

$$T_A + \Sigma U_{A-B} = T_B$$

_____ (2)

Since

$$h = (r_A - r_B) \tan 60° = (60 - 57) \tan 60° = 5.20$$

Solving,

$$v_B = 21.9 \text{ m/s}; \quad \theta = 20.9° \qquad \qquad Ans.$$

15-22. The boy has a weight of 80 lb. While holding on to a ring he runs in a circle and then lifts his feet off the ground, holding himself in the crouched position shown. If *initially* his center of gravity G is $r_A = 10$ ft from the pole and his velocity is *horizontal* such that $v_A = 8$ ft/s, determine (a) his velocity when he is at B, where $r_B = 7$ ft, $\Delta z = 2$ ft, and (b) the vertical component of his velocity, $(v_B)_z$, which is causing him to fall downward.

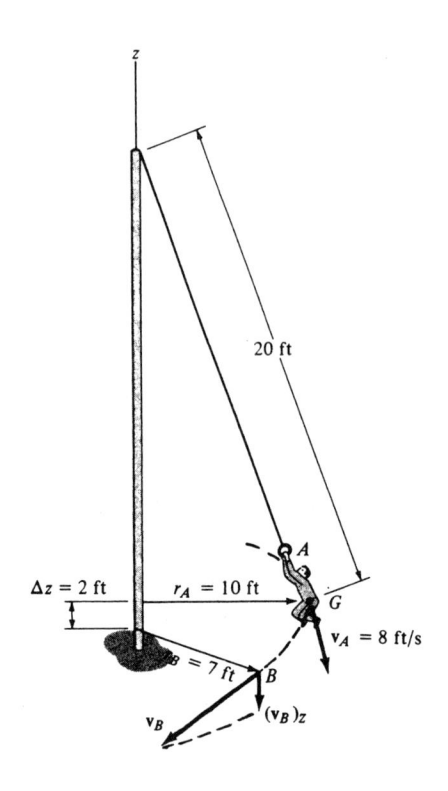

Solution

$$T_A + V_A = T_B + V_B$$

$$v_B = 13.9 \text{ ft/s} \qquad \text{Ans.}$$

$$(H_z)_1 = (H_z)_2$$

$(v_B)_{\text{horiz.}} = 11.43$

$(v_B)_z = \sqrt{v_B^2 - (v_B)_{\text{horiz.}}^2}$

$(v_B)_z = \sqrt{(13.9)^2 - (11.43)^2}$

$(v_B)_z = 7.89$ ft/s

16 Planar Kinematics of a Rigid Body

Rotation About a Fixed Axis

16-1. Gear A is in mesh with gear B as shown. If A starts from rest and has a constant angular acceleration of $\alpha_A = 2$ rad/s^2, determine the time needed for B to attain an angular velocity of $\omega_B = 50$ rad/s.

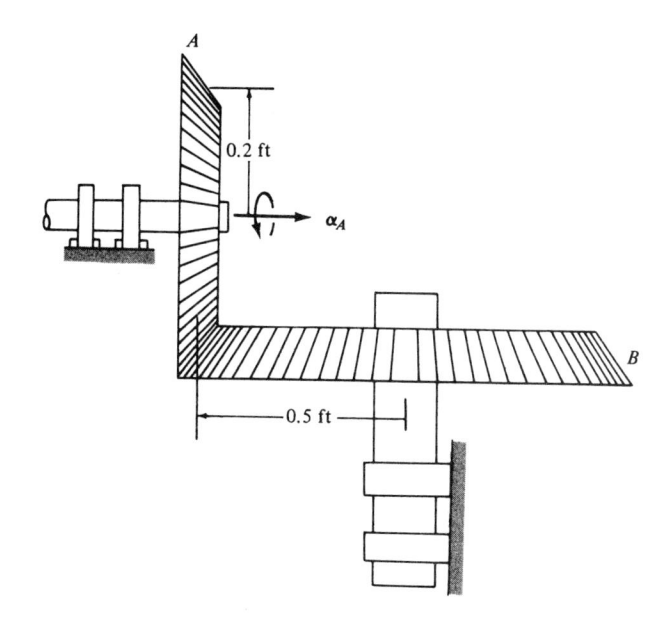

Solution

The point in contact with both gears has a speed of

$$v_P = \omega_B r_B = \underline{\qquad\qquad\qquad\qquad}$$
$$v_P = 25 \text{ ft/s}$$

Thus,

$$\omega_A = \underline{\qquad\qquad\qquad\qquad}$$
$$\omega_A = 125 \text{ rad/s}$$

So that the time is

$$(\omega_A)_2 = (\omega_A)_1 + \alpha_A t$$

$$\underline{\qquad\qquad\qquad\qquad}$$

$$t = 62.5 \text{ s} \qquad\qquad\qquad\qquad Ans.$$

16-2. During a gust of wind, the blades of the windmill are given an angular acceleration of $\alpha = (0.2\,\theta)$ rad/s², where θ is measured in radians. If initially the blades have an angular velocity of 5 rad/s, determine the speed of point P located at the tip of one of the blades just after the blade has turned two revolutions.

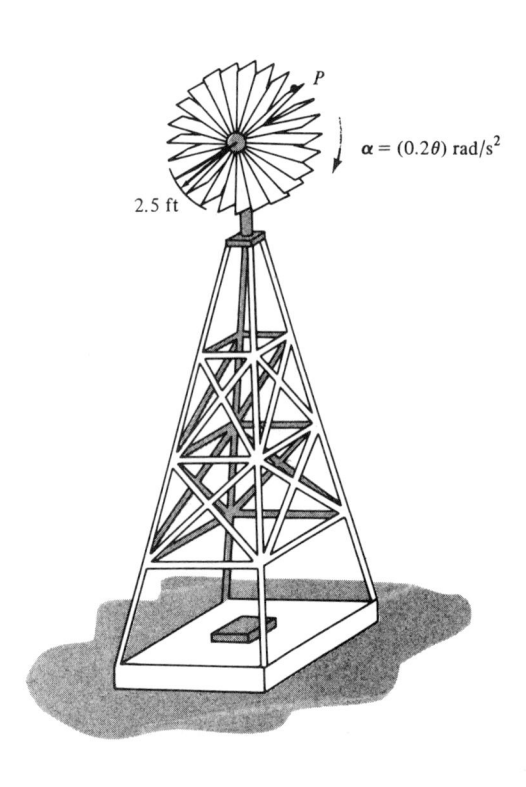

Solution

$$\theta_2 = 2\text{ rev} = 2\pi(2) = 4\pi \text{ rad}$$

$$\omega\,d\omega = \alpha\,d\theta$$

_____

$$\omega = 7.52 \text{ rad/s}$$

$v_P = $ _____

$$v_P = 18.8 \text{ ft/s} \qquad\qquad Ans.$$

16-3. Arm *ABCD* is pinned at *B* and undergoes reciprocating motion such that $\theta = (0.3 \sin 4t)$ rad, where t is measured in seconds and the argument for the sine is in degrees. Determine the largest speed of point *A* during the motion and the magnitude of the acceleration of point *D* at this instant.

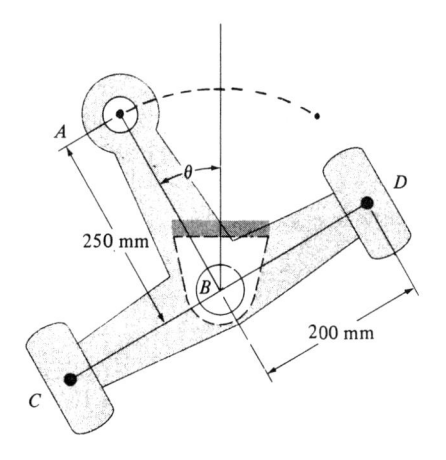

Solution

$$\theta = 0.3 \sin 4t$$
$$\omega = 1.2 \cos 4t$$
$$\omega_{max} = 1.2 \text{ rad/s}$$
$$\alpha = -4.8 \sin 4t$$

ω_{max} occurs when $\cos 4t = 1$ so that $\sin 4t = 0$. Thus $\alpha = 0$

$(v_A)_{max} = $ _____

$(v_A)_{max} = 300$ mm/s   Ans.

$a_A = a_n = $ _____

$a_A = 288$ mm/s²   Ans.

16-4. At the instant shown, gear A is rotating with a constant angular velocity of $\omega_A = 6$ rad/s. Determine the largest angular velocity of gear B and the maximum speed of point C.

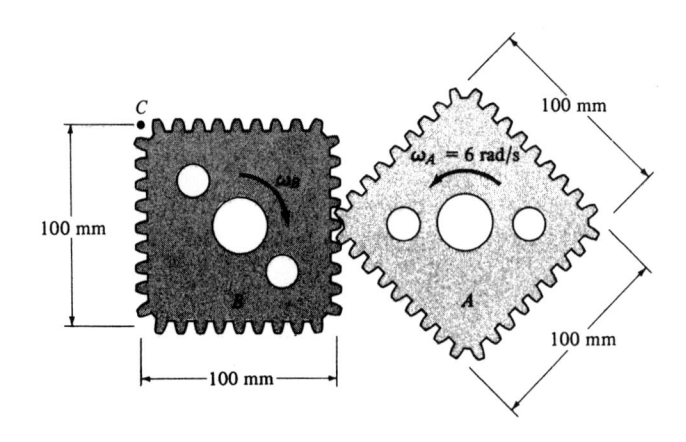

Solution

$$(r_B)_{max} = (r_A)_{max} = 50\sqrt{2} \text{ mm}$$

$$(r_B)_{min} = (r_A)_{min} = 50 \text{ mm}$$

When r_A is max., r_B is min.

Thus

$$\omega_B r_B = \omega_A r_A$$

$(\omega_B)_{max} = $ _____

$(\omega_B)_{max} = 8.49$ rad/s *Ans.*

$v_C = (\omega_B)_{max} r_C = $ _____

$v_C = 0.6$ m/s *Ans.*

16-5. The sphere starts from rest at $\theta = 0$ and rotates with an angular acceleration of $\alpha = (4\theta)$ rad/s², where θ is measured in radians. Determine the magnitudes of the velocity and acceleration of point P on the sphere at the instant $\theta = 6$ rad.

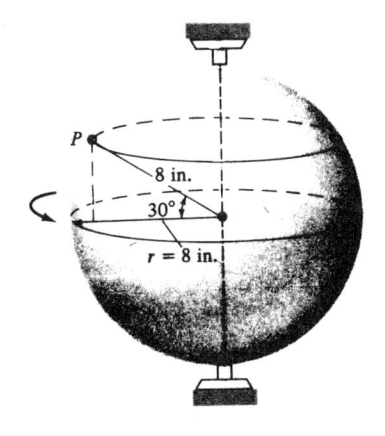

Solution

$$\alpha = 4\theta$$

Since $\omega d\omega = \alpha d\theta$, set up the integrals to determine $\omega = f(\theta)$.

$$\omega = f(\theta)$$

$$\omega = 2\theta$$

When $\theta = 6$ rad

$$\alpha = 24 \text{ rad/s}^2, \quad \omega = 12 \text{ rad/s}$$

$$v_P = \omega r' =$$

$$v_P = 83.14 \text{ in/s} = 6.93 \text{ ft/s} \qquad \textit{Ans.}$$

$$a_n = \frac{v^2}{r'} = \frac{(83.14)^2}{(8 \cos 30°)} = 997.7$$

$$a_t = \alpha r' = 24(8 \cos 30°) = 166.3$$

$$a_P = \sqrt{(997.7)^2 + (166.3)^2}$$

$$a_P = 1011 \text{ in/s}^2 = 84.3 \text{ ft/s} \qquad \textit{Ans.}$$

Absolute General Plane Motion Analysis

16-6. The mechanism is used to convert the *constant* circular motion of rod *AB* into translating motion of rod *CD*. Compute the velocity and acceleration of *CD* for any angle θ of *AB*.

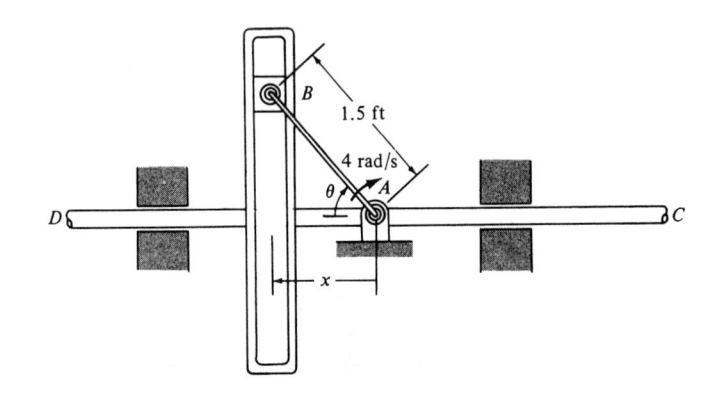

Solution

$x =$ _____

$\dot{x} =$ _____

$\ddot{x} =$ _____

$$v_{CD} = -6 \sin\theta \text{ ft/s} \qquad \qquad Ans.$$

Since

$$\ddot{\theta} = 0$$

$$a_{CD} = -24 \cos\theta \text{ ft/s}^2 \qquad \qquad Ans.$$

16-7. Rod *CD* presses against *AB*, giving it an angular velocity. If the angular velocity of *AB* is maintained at $\omega = 5$ rad/s, determine the required speed **v** of *CD* for any angle θ of rod *AB*.

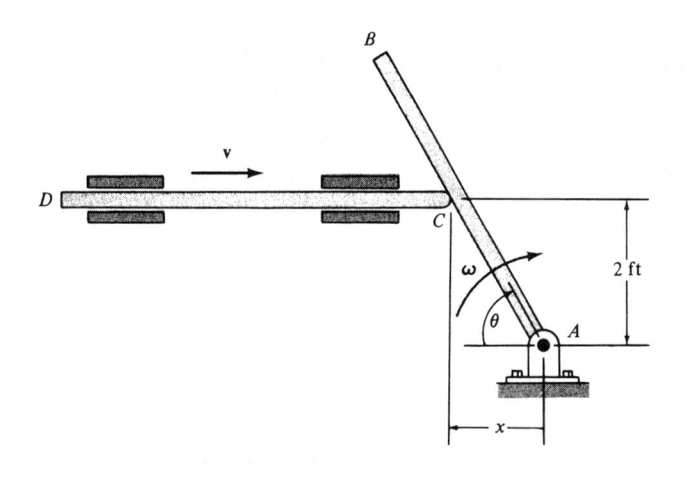

Solution

$x = $ _____

$\dot{x} = $ _____

$v = -10 \csc^2 \theta$ *Ans.*

16-8. The safe is transported on a platform which rests on rollers, each having a radius r. If the rollers do not slip, determine their angular velocity ω if the safe moves forward with a velocity v.

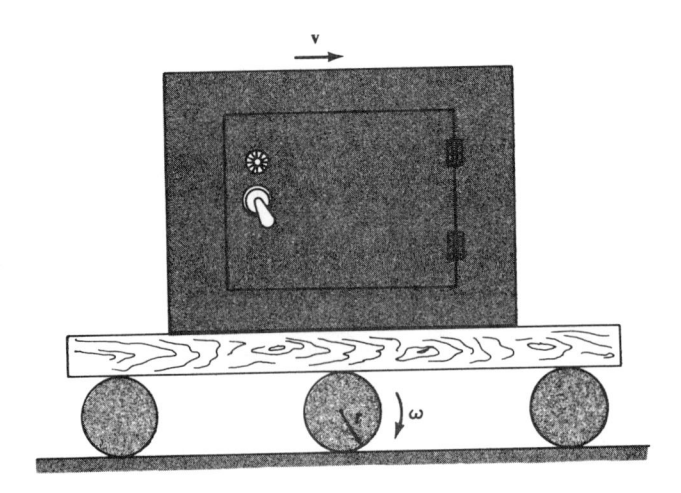

Solution

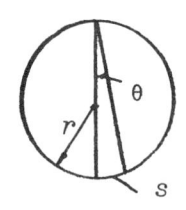

From the figure

$$s = \underline{\hspace{3in}}$$

$$v = \underline{\hspace{3in}}$$

So that

$$\omega = v/2r \qquad\qquad Ans.$$

16-9. The scaffold S is raised hydraulically by moving the roller at A towards the pin at B. If A is approaching B with a speed of 1.5 ft/s, determine the speed at which the platform is rising as a function of θ. Each link is pin-connected at its midpoint and end points and has a length of 4 ft.

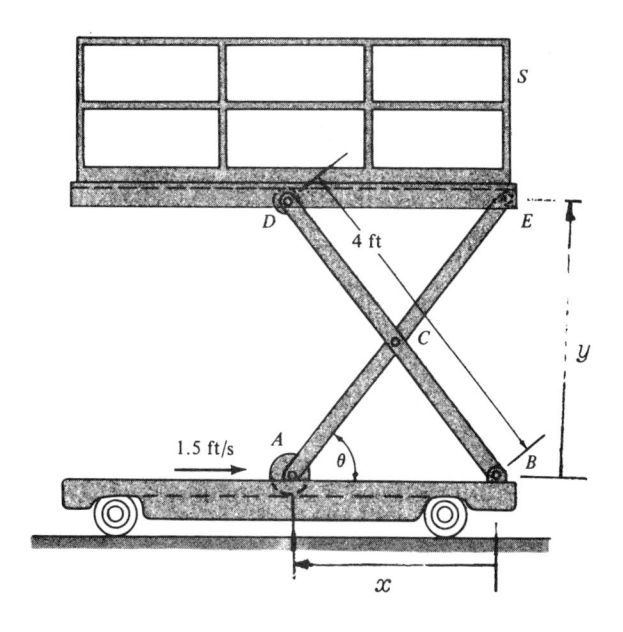

Solution

$$y = \underline{\hspace{4in}}$$

$$\dot{y} = \underline{\hspace{4in}} \quad (1)$$

$$x = \underline{\hspace{4in}}$$

$$\dot{x} = \underline{\hspace{4in}} \quad (2)$$

Eliminating $\dot{\theta}$ between Eqs. (1) and (2) yields

$$\dot{y} = -\dot{x} \cot\theta$$

$$v_S = -v_A \cot\theta$$

$$v_S = 1.5 \cot\theta \text{ ft/s} \qquad \textit{Ans.}$$

16-10. The 2-m-long bar is confined to move in the horizontal and vertical slots A and B. If the velocity of the slider block at A is 6 m/s, determine the bar's angular velocity and the velocity of block B at the instant $\theta = 60°$.

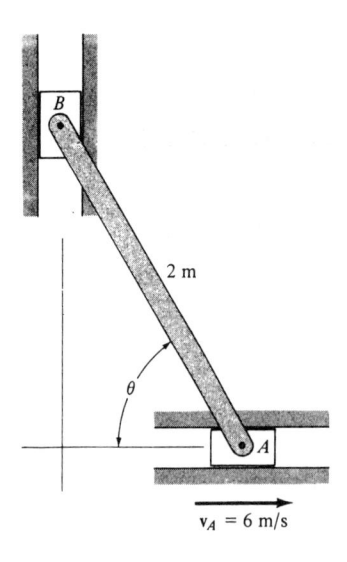

Solution

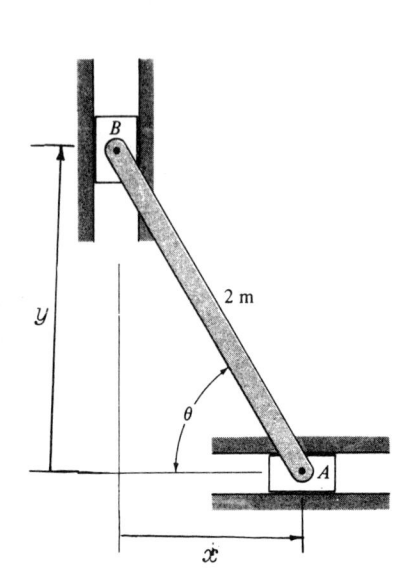

$x =$ _____

$v_A =$ _____

When $v_A = 6$ m/s, $\theta = 60°$,

$\omega = 3.46$ rad/s $\curvearrowleft$ Ans.

$y =$ _____

$v_B =$ _____

When

$\omega = 3.46$ rad/s, $\theta = 60°$,

$v_B = 3.46$ rad/s $\downarrow$ Ans.

Relative-Motion Analysis: Velocity

16-11. Due to an engine failure, the missile is rotating at $\omega = 3$ rad/s, while its mass center G is moving upward at 200 ft/s. Determine the velocity of its nose B at this instant.

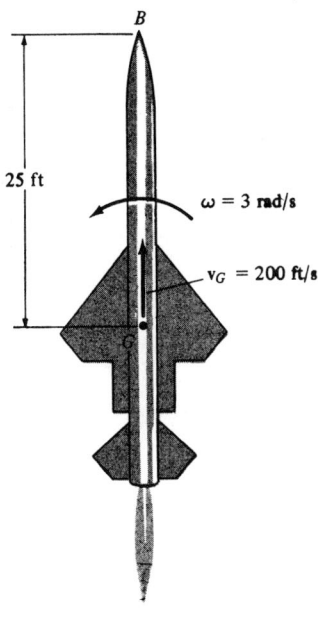

Solution

Apply the relative velocity equation

$$\mathbf{v}_B = \mathbf{v}_G + \boldsymbol{\omega} \times \mathbf{r}_{B/G}$$

$(v_B)_x \mathbf{i} + (v_B)_y \mathbf{j} = $ _____

$(v_B)_x = -3(25)$

$(v_B)_y = 200$

$v_B = \sqrt{(200)^2 + (75)^2}$

$v_B = 214$ ft/s *Ans.*

$\theta = \tan^{-1}\left(\dfrac{200}{75}\right) = 69.4°$ *Ans.*

16-12. If the block at C is moving downward at 4 ft/s, determine the angular velocity of bar AB at the instant shown.

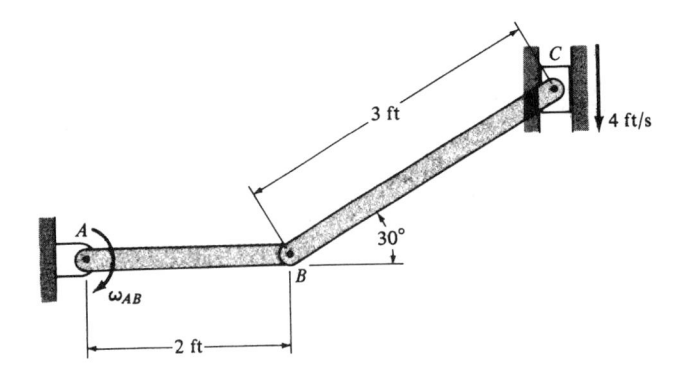

Solution

Apply the relative velocity equation

$$\mathbf{v}_B = \mathbf{v}_C + \boldsymbol{\omega}_{BC} \times \mathbf{r}_{B/C}$$

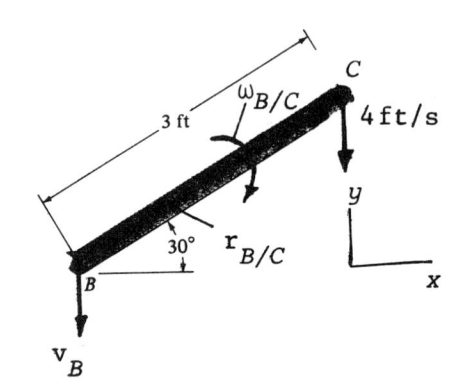

$$0 = -3\omega_{BC} \sin 30°$$

$$-\omega_{AB}(2) = -4 + 3\omega_{BC} \cos 30°$$

Solving:

$$\omega_{BC} = 0$$

$$\omega_{AB} = 2 \text{ rad/s} \qquad \text{Ans.}$$

16-13. Knowing the angular velocity of link CD is $\omega_{CD} = 4$ rad/s, determine the angular velocities of links BC and AB at the instant shown.

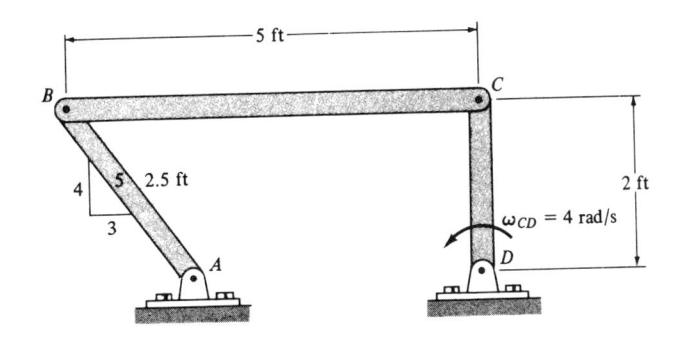

Solution

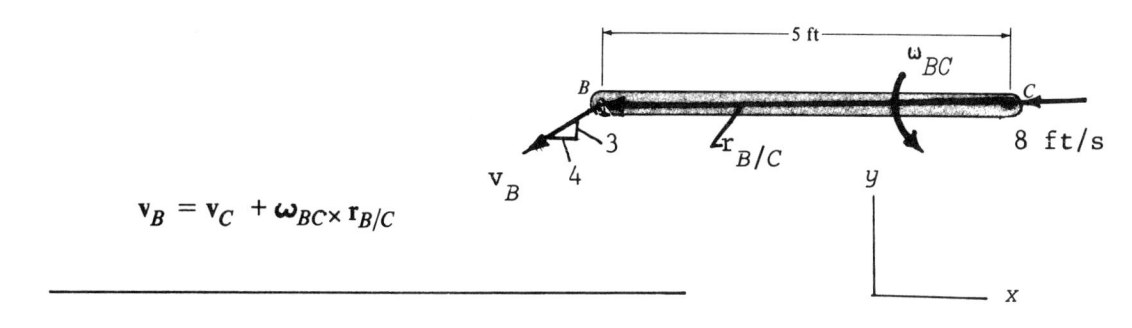

$$\mathbf{v}_B = \mathbf{v}_C + \boldsymbol{\omega}_{BC} \times \mathbf{r}_{B/C}$$

$$-\left(\frac{4}{5}\right) v_B = -8$$

$$-\left(\frac{3}{5}\right) v_B = -5\omega_{BC}$$

Solving

$$v_B = 10 \text{ ft/s}$$

$$\omega_{BC} = 1.20 \text{ ft/s} \qquad \qquad Ans.$$

$$v_B = r_{AB}\omega_{AB}$$

$$10 = 2.5\omega_{AB}$$

$$\omega_{AB} = 4 \text{ rad/s} \qquad \qquad Ans.$$

16-14. If rod CD has a downward velocity of 6 in./s at the instant shown, determine the velocity of the gear rack A at this instant. The rod is pinned at C to gear B.

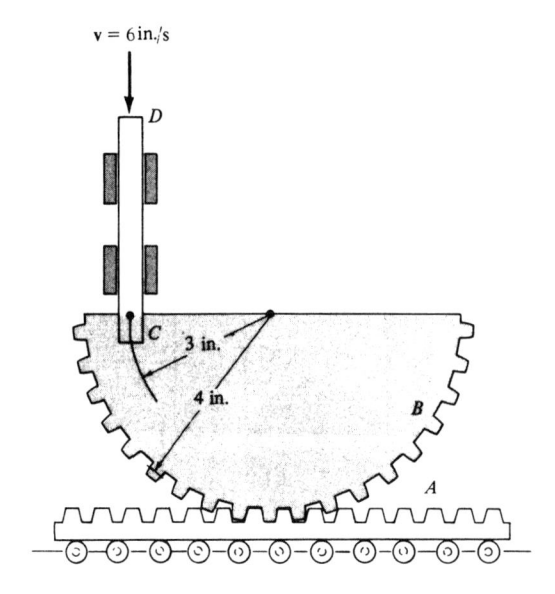

Solution

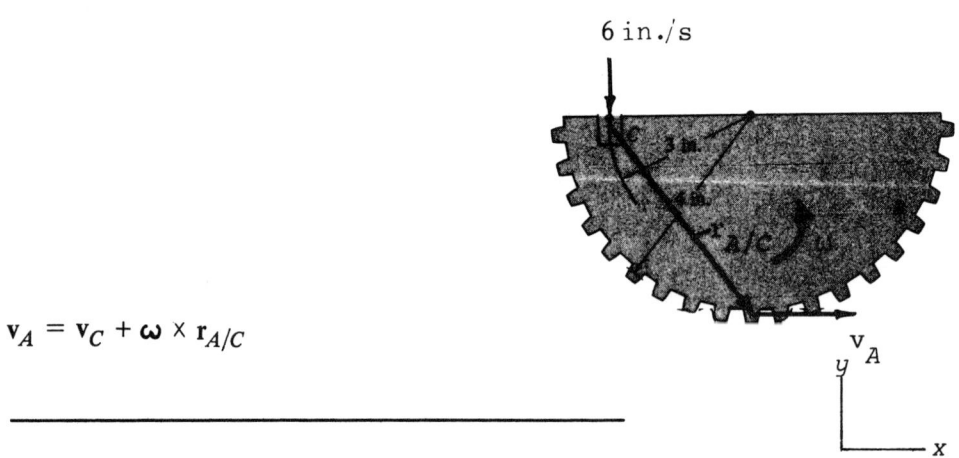

$$\mathbf{v}_A = \mathbf{v}_C + \boldsymbol{\omega} \times \mathbf{r}_{A/C}$$

$v_A = 4\omega$

$0 = -6 + 3\omega$

Solving,

$\omega = 24 \text{ rad/s}$

$v_A = 8 \text{ in./s}$ *Ans.*

16-15. The rotation of link AB creates an oscillating movement of gear F. If AB has an angular velocity of $\omega_{AB} = 8$ rad/s, determine the angular velocity of gear F at the instant shown. Gear E is a part of arm CD and pinned at D to a fixed point.

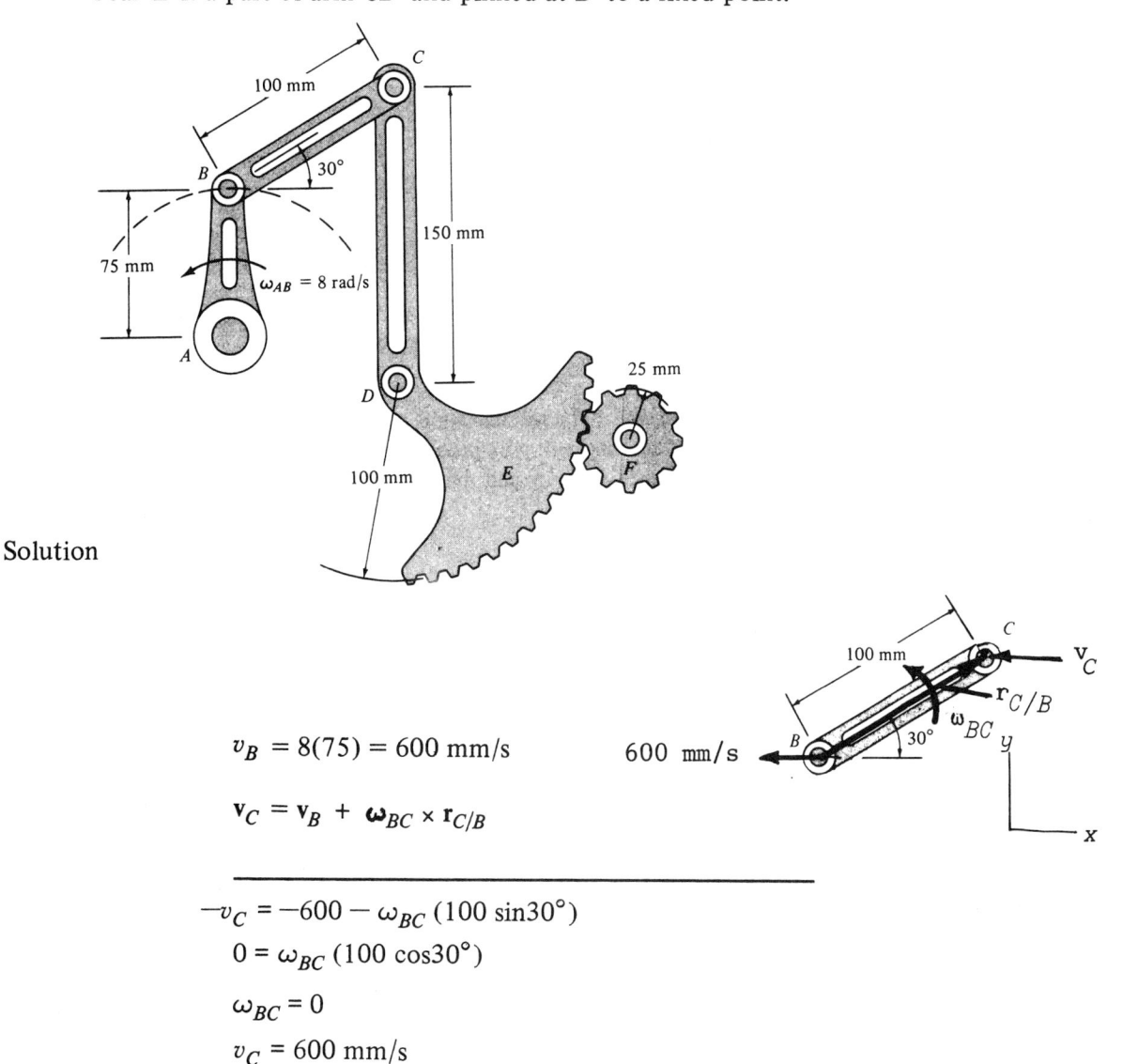

Solution

$$v_B = 8(75) = 600 \text{ mm/s}$$

$$\mathbf{v}_C = \mathbf{v}_B + \boldsymbol{\omega}_{BC} \times \mathbf{r}_{C/B}$$

$$-v_C = -600 - \omega_{BC}(100 \sin 30°)$$
$$0 = \omega_{BC}(100 \cos 30°)$$
$$\omega_{BC} = 0$$
$$v_C = 600 \text{ mm/s}$$

Since E rotates about point D,

$$v_C = (r_{C/D})(\omega_E)$$
$$600 = 150(\omega_E)$$
$$\omega_E = 4 \text{ rad/s}$$

The speed of a point P in contact between gears E and F is

$$v_P = \omega_E r_{DE} = 4(100)$$
$$v_P = 400 \text{ mm/s}$$
$$\omega_F = \frac{400}{25}$$
$$\omega_F = 16 \text{ rad/s}$$

Instantaneous Center of Zero Velocity

16-16. The automobile with wheels 2.5 ft in diameter is traveling in a straight path at a rate of 60 ft/s. If no slipping occurs, determine the angular velocity of one of the rear wheels and the velocity of the fastest moving point on the wheel.

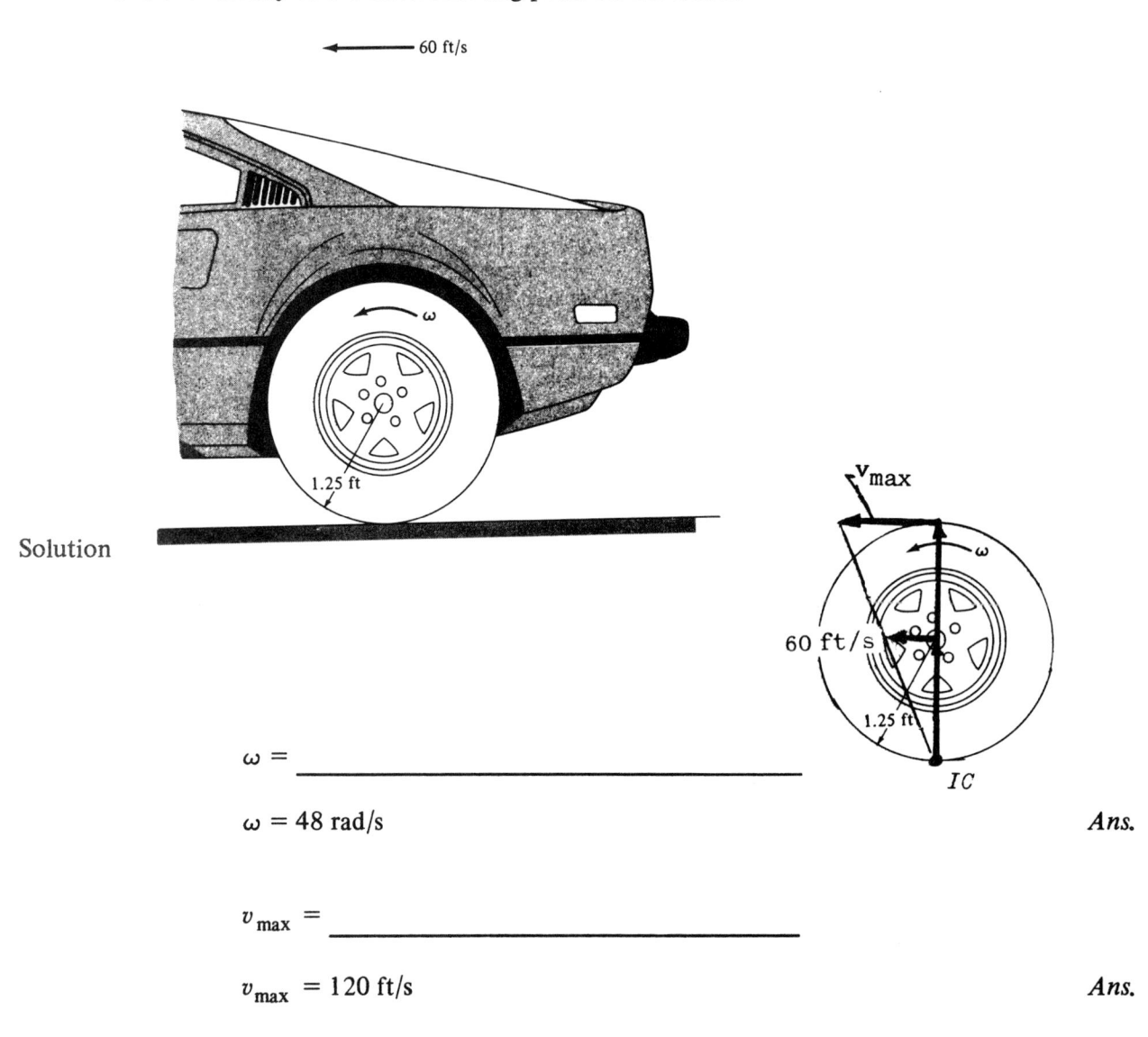

Solution

$\omega =$ _____

$\omega = 48$ rad/s Ans.

$v_{max} =$ _____

$v_{max} = 120$ ft/s Ans.

16-17. As the cord unravels from the wheel's inner hub, the wheel is rotating at $\omega = 2$ rad/s at the instant shown. Determine the velocities of points A and B.

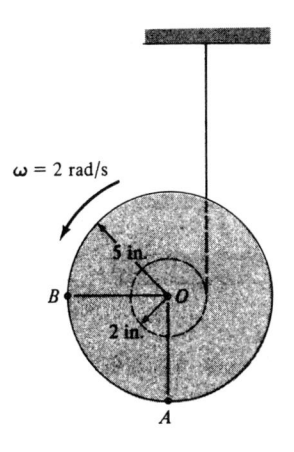

Solution

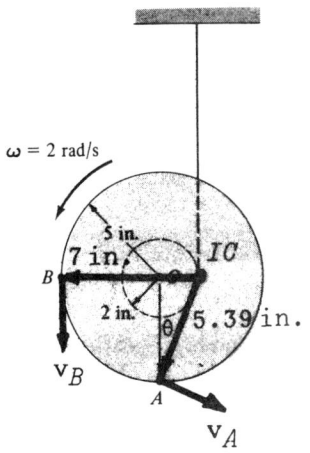

$v_B = $ _____

$\quad\quad = 14$ in./s↓ Ans.

$v_A = $ _____

$v_A = 10.8$ in./s Ans.

$\theta = \tan^{-1}$ _____

$\theta = 21.8°$ Ans.

16-18. Part of an automatic transmission consists of a *fixed* ring gear R, three equal planet gears P, the sun gear S, and the planet carrier C, which is shaded. If the sun gear is rotating at $\omega_S = 5$ rad/s, determine the angular velocity of the *planet carrier*. Note that C is pin-connected to the center of each of the planet gears.

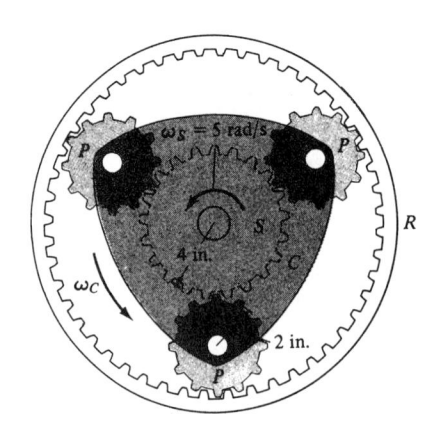

Solution

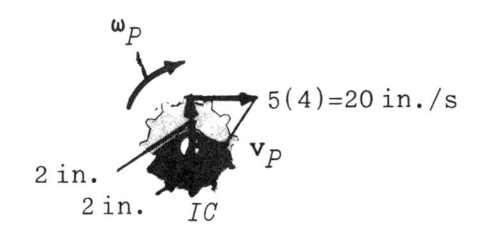

$\omega_P = 5$ rad/s

$v_P = $ _____

$v_P = 10$ in./s

$\omega_C = $ _____

$\omega_C = 1.66$ rad/s   Ans.

16-19. Show that if the rim of the wheel and its hub maintain contact with the three tracks as the wheel rolls, it is necessary that slipping occurs at the hub A if no slipping occurs at B. Under these conditions, what is the speed at A if the wheel has an angular velocity ω?

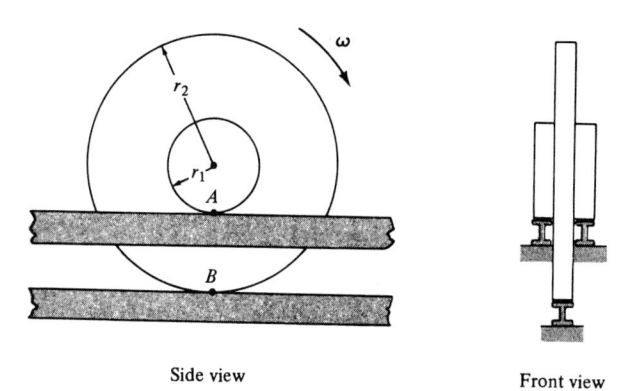

Side view

Front view

Solution

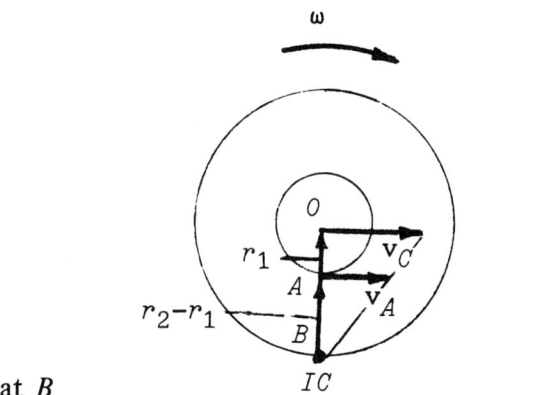

IC is at B

$v_A = $ _____ *Ans.*

$v_C = $ _____

∴ Slipping occurs at A since $v_A \neq 0$

16-20. The oil pumping unit consists of a walking beam AB, connecting rod BC, and crank CD. If the crank rotates at a constant rate of 6 rad/s, determine the speed of the rod hanger H at the instant shown. *Hint*: Point B follows a circular path about point E and therefore the velocity of B is *not* vertical.

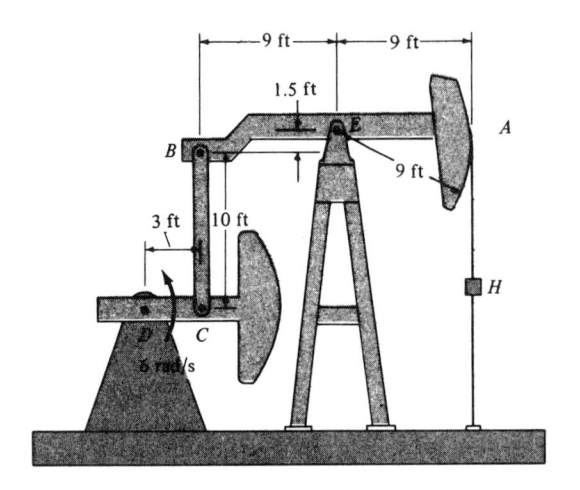

Solution

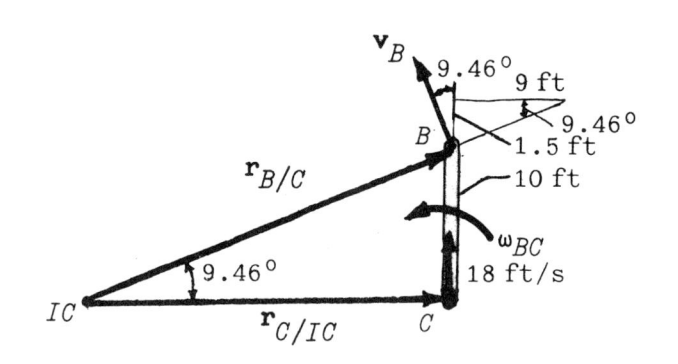

$r_{C/IC} = $ _____

$r_{C/IC} = 60$ ft

$\omega_{BC} = $ _____

$\omega_{BC} = 0.300$ rad/s

$v_B = $ _____

$v_B = 18.25$ ft/s

$\omega_{BA} = 18.25/\sqrt{(9)^2 + (1.5)^2}$

$\omega_{BA} = 2.0$ rad/s

$v_H = v_A = 2(9)$

$v_H = 18$ ft/s   Ans.

Relative-Motion Analysis: Acceleration

16-21. Determine the angular acceleration of link BC at the instant $\theta = 90°$ if the collar C has an instantaneous velocity of $v_C = 4$ ft/s and deceleration of $a_C = 3$ ft/s² as shown.

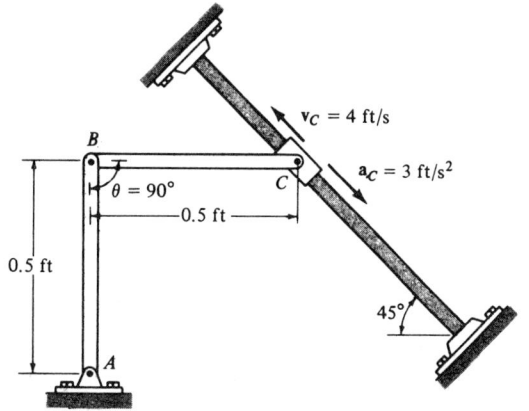

Solution

Velocity Analysis

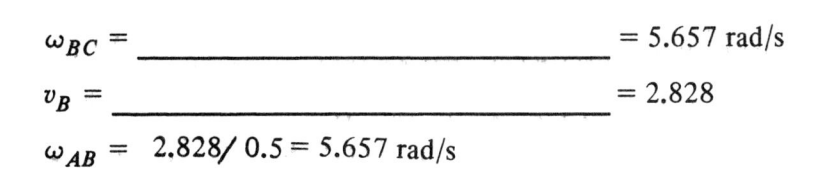

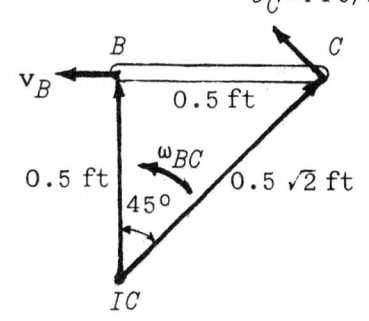

$\omega_{BC} = \underline{\hspace{4cm}} = 5.657$ rad/s

$v_B = \underline{\hspace{4cm}} = 2.828$

$\omega_{AB} = 2.828 / 0.5 = 5.657$ rad/s

Acceleration Analysis

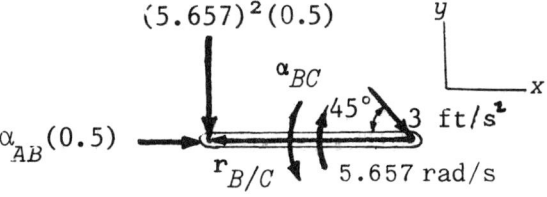

$(a_B)_x + (a_B)_y = a_C + \alpha_{BC} \times r_{B/C} - \omega_{BC}^2 r_{B/C}$

$(a_B)_x = 3 \cos 45° + (5.657)^2 (0.5)$

$-(5.657)^2 (0.5) = -3 \sin 45° - \alpha_{BC}(0.5)$

Solving

$(a_B)_x = 18.1$ ft/s²

$\alpha_{BC} = 27.8$ rad/s² $\;\;$ Ans.

16-22. The pulley is pin-connected to block B at A. As cord CF unwinds from the inner hub with the motion shown, cord DE unwinds from the outer rim. Determine the angular acceleration of the pulley at the instant shown.

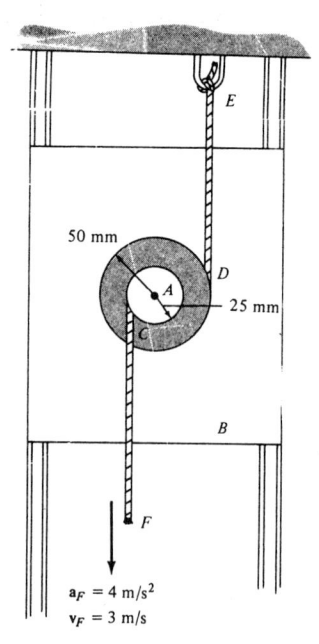

Solution
Velocity Analysis

$$\omega = \frac{3}{0.075} = 40 \text{ rad/s}$$

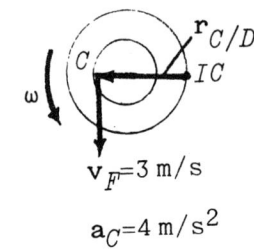

Acceleration Analysis

$$\mathbf{a}_C = \mathbf{a}_D + \boldsymbol{\alpha} \times \mathbf{r}_{C/D} - \omega^2 \, \mathbf{r}_{C/D}$$

$(a_C)_x = (40)^2 (0.075)$

$-4 = -\alpha (0.075)$

$\alpha = 53.3 \text{ rad/s}^2$ Ans.

16-23. The disk rolls without slipping such that it has an angular acceleration of $\alpha = 4$ rad/s² and angular velocity of $\omega = 2$ rad/s at the instant shown. Determine the accelerations of points A and B on the link and the link's angular acceleration at this instant. Assume point A lies on the periphery of the disk, 150 mm from C.

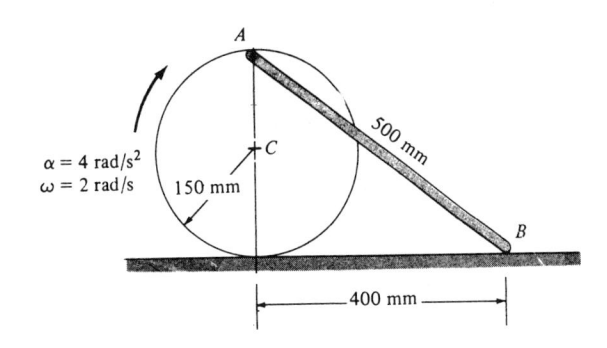

Solution

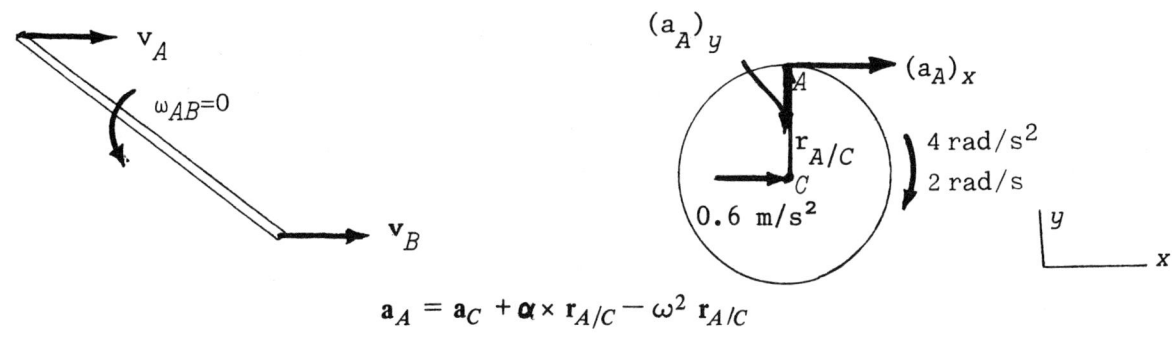

$$\mathbf{a}_A = \mathbf{a}_C + \boldsymbol{\alpha} \times \mathbf{r}_{A/C} - \omega^2 \mathbf{r}_{A/C}$$

$(a_A)_x = 0.6 + 4(0.15)$

$(a_A)_y = -(2)^2 (0.15)$

$(a_A)_x = 1.20$ m/s² →   Ans.

$(a_A)_y = 0.6$ m/s² ↓   Ans.

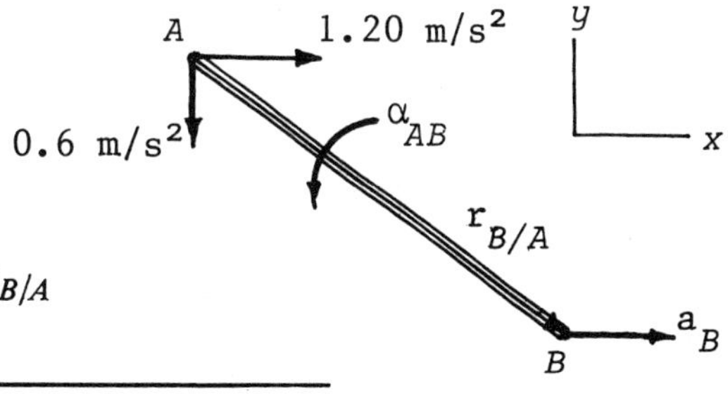

$$\mathbf{a}_B = \mathbf{a}_A + \boldsymbol{\alpha}_{AB} \times \mathbf{r}_{B/A} - \omega_{AB}^2 \mathbf{r}_{B/A}$$

$a_B = 1.20 + 0.3\, \alpha_{AB}$

$0 = -0.6 + 0.4\, \alpha_{AB}$

$\alpha_{AB} = 1.50$ rad/s²   Ans.

$a_B = 1.65$ m/s²   Ans.

16-24. Gear C is rotating with a constant angular velocity of $\omega_C = 3$ rad/s. Determine the acceleration of the piston A and the angular acceleration of rod AB at the instant $\theta = 90°$. Set $r_C = 0.2$ ft and $r_D = 0.3$ ft.

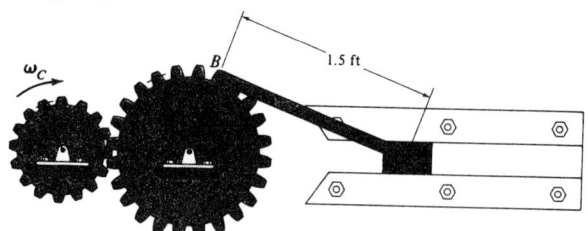

Solution

Velocity Analysis

$v_P = \omega_C r_C = (3)(0.2) = 0.6$ ft/s↓

$\omega_D = \dfrac{v_P}{r_D} = \dfrac{0.6}{0.3} = 2$ rad/s ⟳

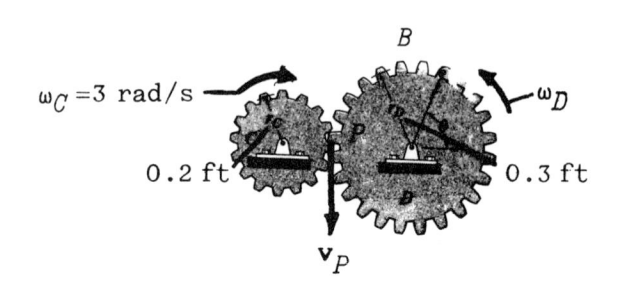

$v_B = v_P = 0.6$ ft/s ←

Since the IC is at infinity, the rod AB is in translation at the instant considered.

Acceleration Analysis

Since $\alpha_C = 0$, then $\alpha_D = 0$

$\mathbf{a}_B = (\mathbf{a}_B)_t + (\mathbf{a}_B)_n = 0\mathbf{i} - (2)^2 (0.3)\mathbf{j} = -1.20\mathbf{j}$

$\mathbf{a}_B = \mathbf{a}_A + \boldsymbol{\alpha}_{AB} \times \mathbf{r}_{B/A} - \omega_{AB}^2\, \mathbf{r}_{B/A}$

$0 = a_A - 0.3\, \alpha_{AB}$

$-1.20 = -1.47\, \alpha_{AB}$

Solving,

$a_A = 0.245$ ft/s² →; $\alpha_{AB} = 0.816$ rad/s² ⟳ **Ans.**

16-25. At the instant shown, arm AB has an angular velocity of $\omega_{AB} = 0.5$ rad/s and an angular acceleration of $\alpha_{AB} = 2$ rad/s^2. Determine the angular velocity and angular acceleration of the dump bucket at this instant.

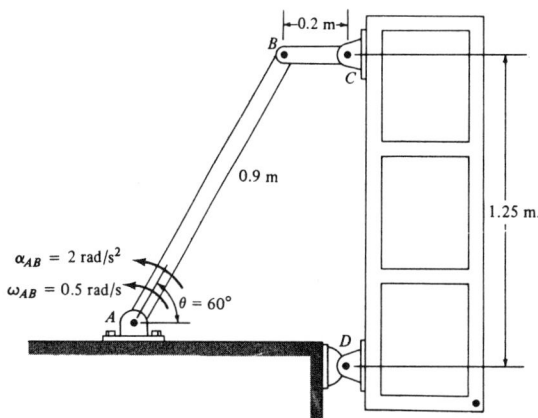

Solution
Velocity Analysis

$v_B = \omega_{AB}\, r_{B/A} = 0.5\,(0.9) = 0.45$ m/s

$$\omega_{BC} = \frac{v_B}{r_{B/IC}} = \frac{0.45}{0.2/\cos 60°} = 1.125 \text{ rad/s}\;\;\rotatebox{0}{)}$$

$v_C = \omega_{BC}\, r_{C/IC} = 1.125\,(0.2)(\tan 60°) = 0.390$ m/s ←

$$\omega_{DC} = \frac{v_C}{r_{C/D}} = \frac{0.390}{1.25} = 0.312 \text{ rad/s} \;\;\rotatebox{0}{)} \qquad Ans.$$

Acceleration Analysis

$(a_B)_t = (2)(0.9) = 1.80$ m/s^2

$(a_B)_n = (0.5)^2 (0.9) = 0.225$ m/s^2

$(a_C)_n = (0.312)^2 (1.25) = 0.122$ m/s

$\mathbf{a}_C = \mathbf{a}_B + \boldsymbol{\alpha}_{BC} \times \mathbf{r}_{C/B} - (\omega_{BC})^2\, \mathbf{r}_{C/B}$

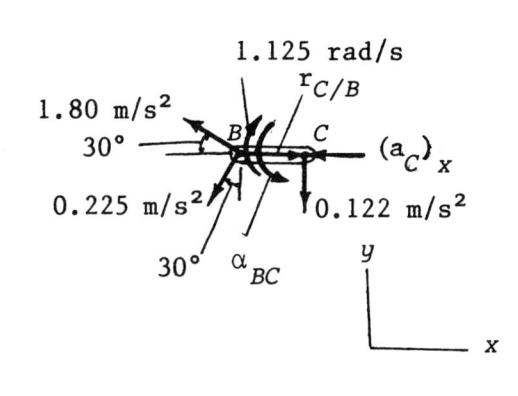

$-(a_C)_x = -1.80 \cos 30° - 0.225 \sin 30° - (1.125)^2 (0.2)$

$-0.122 = 1.80 \sin 30° - 0.225 \cos 30° + \alpha_{BC}(0.2)$

Solving, $(a_C)_x = 1.92$ m/s^2; $\alpha_{BC} = -4.14$ rad/s^2

Thus,

$$\alpha_{DC} = \frac{(a_C)_x}{r_{C/D}} = \frac{1.92}{1.25} = 1.54 \text{ rad/s}^2 \;\;\rotatebox{0}{)} \qquad Ans.$$

17 Planar Kinetics of a Rigid Body: Force and Acceleration

Equations of Motion: Translation

17-1. The sports car has a mass of 1.5 Mg and a center of mass at G. Determine the shortest time it takes for it to reach a speed of 80 km/h, starting from rest, if the engine only drives the rear wheels, whereas the front wheels are free rolling. The coefficient of friction between the wheels and road is $\mu = 0.2$. Neglect the mass of the wheels for the calculation.

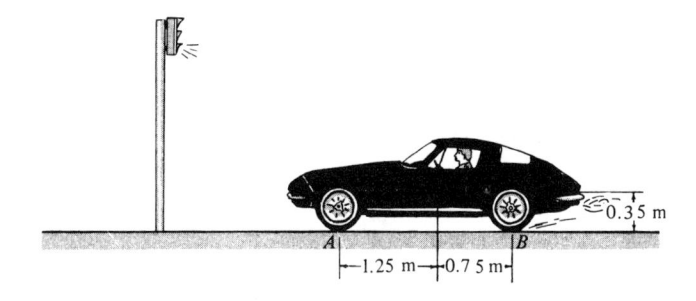

Solution

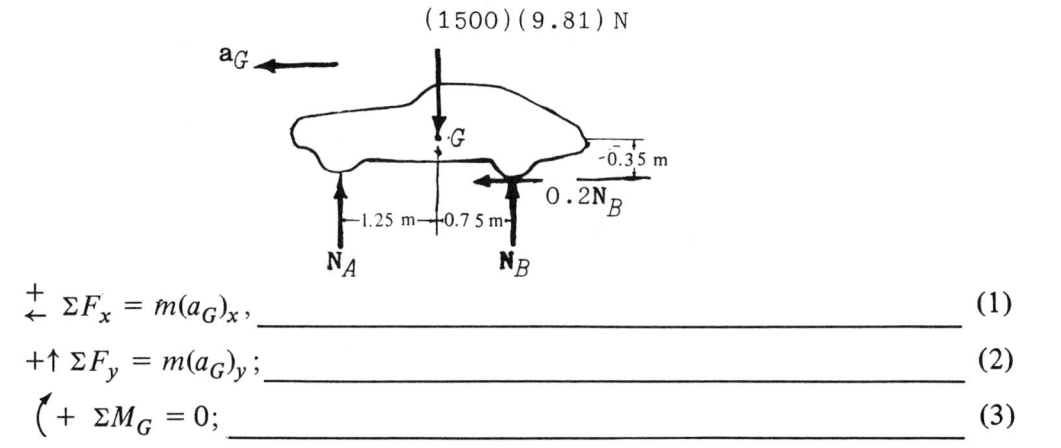

$\xrightarrow{+} \Sigma F_x = m(a_G)_x,$ _____ (1)

$+\uparrow \Sigma F_y = m(a_G)_y;$ _____ (2)

$(+ \Sigma M_G = 0;$ _____ (3)

Solving Eqs. (1) – (3)

$N_A = 5.18$ kN; $N_B = 9.53$ kN; $a_G = 1.27$ m/s²

Since $v_2 = 80$ km/h $= 22.2$ m/s, then

$\xrightarrow{+} v_2 = v_1 + a_G t$

$22.2 = 0 + 1.96t$

$t = 11.3$ s   Ans.

17-2. Bar AB has a weight of 10 lb and is fixed to the carriage at A. Determine the internal axial force A_y, shear force V_x, and moment M_A at A if the carriage is descending the plane with an acceleration of 4 ft/s².

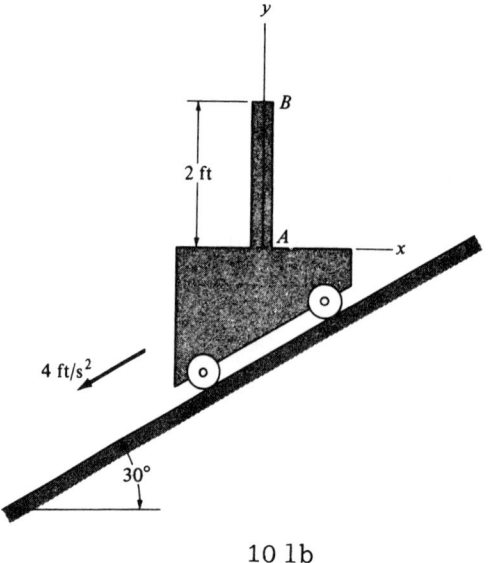

Solution

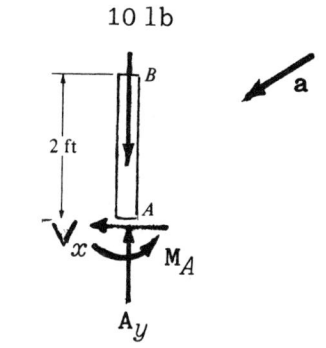

$+\uparrow \Sigma F_y = m(a_G)_y;$ _____

$\qquad\qquad\qquad\qquad A_y = 9.38$ N Ans.

$\xleftarrow{+} \Sigma F_x = m(a_G)_x;$ _____

$\qquad\qquad\qquad\qquad V_x = 1.08$ N Ans.

$(\!\!+\ \Sigma M_A = \Sigma(M_k)_A;$ _____

$\qquad\qquad\qquad\qquad M_A = 1.08$ N·m Ans.

17-3. The dragster has a mass of 1.3 Mg and a center of mass at G. If a braking parachute is attached at C and provides a horizontal braking force of $F = (1.8\,v^2)$ N, where v is in m/s, determine the deceleration the dragster can have upon releasing the parachute such that the wheels at B are on the verge of leaving the ground, i.e., the normal reaction at B is zero. Neglect the mass of the wheels and assume the engine is disengaged so that the wheels are freely rolling.

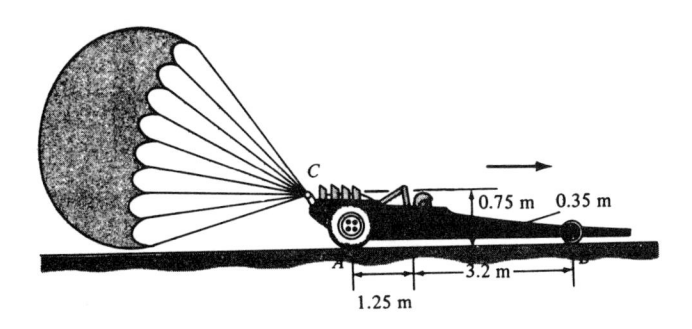

Solution

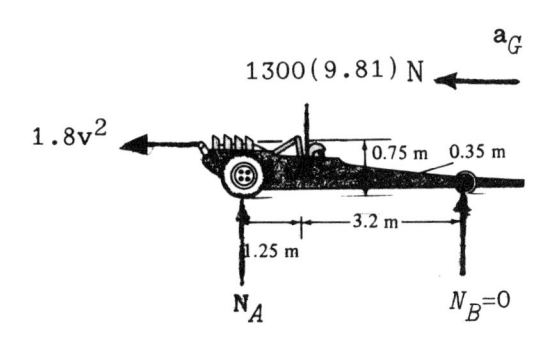

$\xleftarrow{+} \Sigma F_x = m(a_G)_x;$ _____

$\left(+ \Sigma M_G = 0;\right.$ _____

$+\uparrow \Sigma F_y = m(a_G)_y;$ _____

Solving,
$$a_G = 16.4 \text{ m/s}^2 \qquad\qquad Ans.$$

17-4. The 10-kg block rests on the platform for which $\mu = 0.4$. If at the instant shown link AB has an angular velocity $\omega = 2$ rad/s, determine the greatest angular acceleration of the link so that the block doesn't slip.

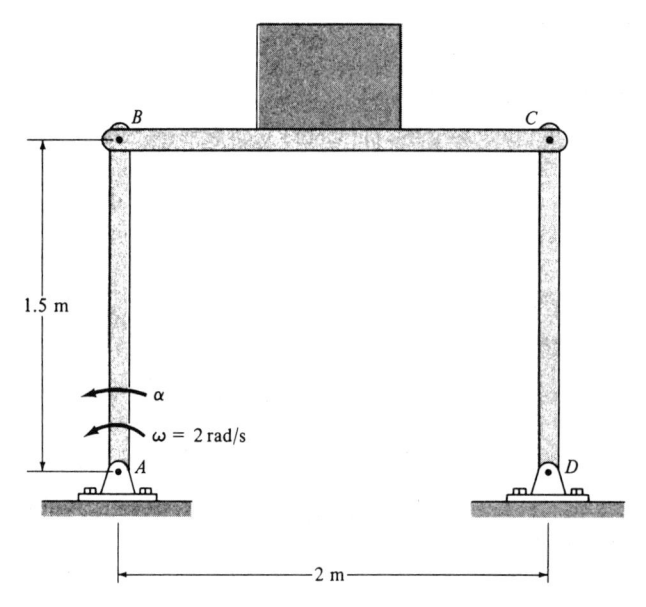

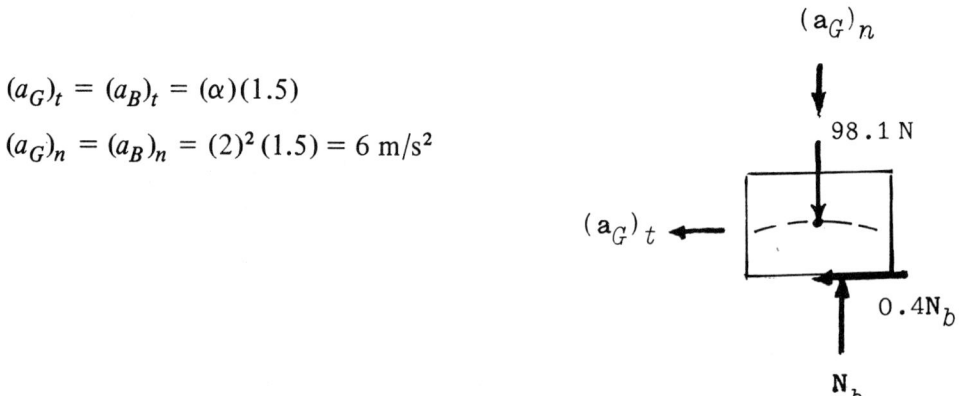

Solution

$(a_G)_t = (a_B)_t = (\alpha)(1.5)$

$(a_G)_n = (a_B)_n = (2)^2(1.5) = 6 \text{ m/s}^2$

$+\uparrow \Sigma F_n = m(a_G)_n$

$$N_b = 38.1 \text{ N}$$

$\xleftarrow{+} \Sigma F_t = m(a_G)_t$

$$\alpha = 1.02 \text{ rad/s}^2 \qquad \textit{Ans.}$$

17-5. The crate C has a weight of 150 lb and rests on the floor of a truck elevator for which $\mu = 0.4$. Determine the largest initial angular acceleration α, starting from rest, which the parallel links AB and DE can have without causing the crate to slip. No tipping occurs.

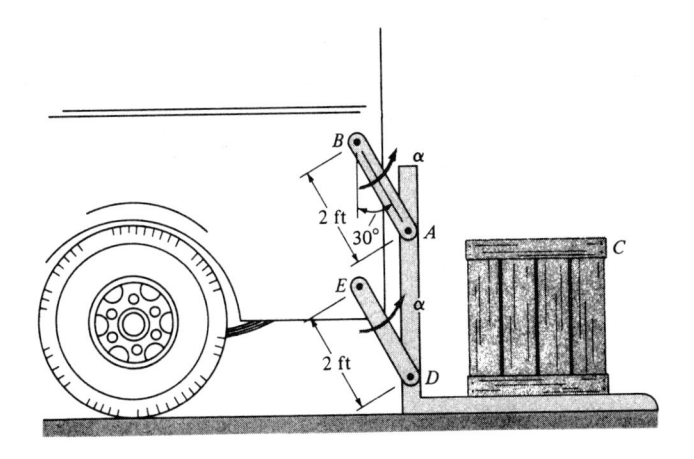

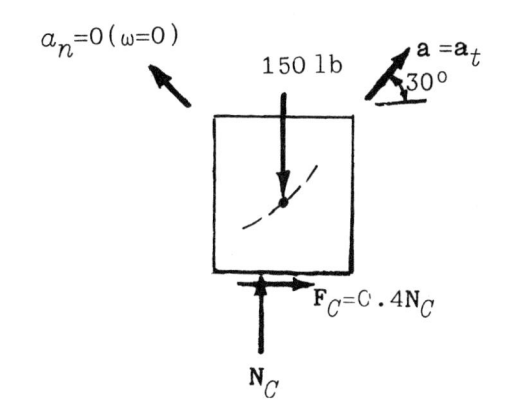

Solution

$\xrightarrow{+} \Sigma F_x = ma_x;$ _____

$+\uparrow \Sigma F_y = ma_y;$ _____

Solving $N_C = 195.0$ lb

$a = 19.34$ ft/s²

$19.34 = (2)\alpha$

$\alpha = 9.67$ rad/s² *Ans.*

17-6. The trailer portion of a truck has a mass of 4 Mg with a center of mass at G. If a *uniform* crate, having a mass of 800 kg and a center of mass at G_c, rests on the trailer, determine the horizontal and vertical components of reaction at the ball-and-socket joint (pin) A when the truck is decelerating at a constant rate of $a = 3$ m/s². Assume that the crate does not slip on the trailer and neglect the mass of the wheels. The wheels at B roll freely.

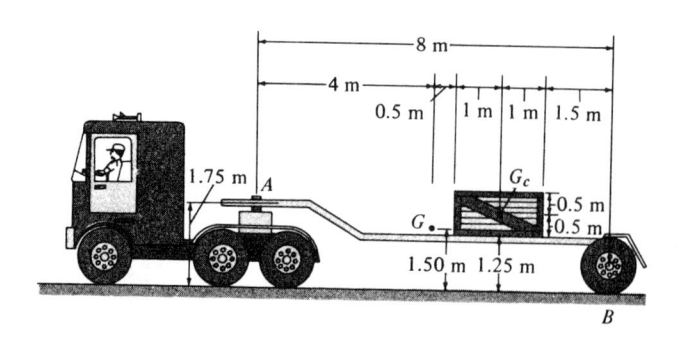

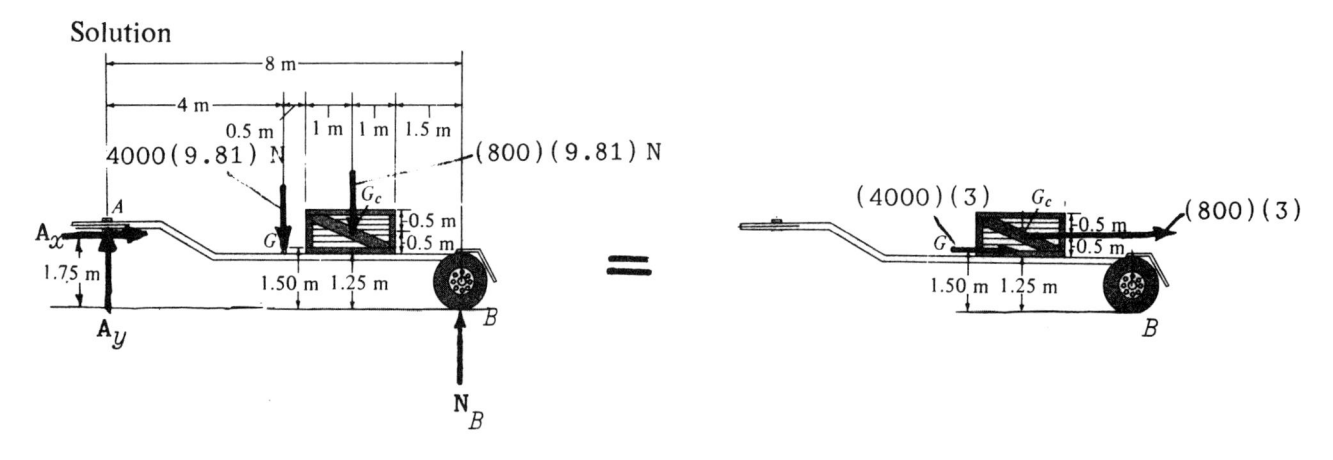

Solution

$\xrightarrow{+} \Sigma F_x = m(a_G)_x;$ _____

$$A_x = 14\,400 \text{ N} = 14.4 \text{ kN} \qquad\qquad Ans.$$

$(\!\!+\ \Sigma M_B = \Sigma(M_k)_B;$ _____

$$A_y = 21\,697.5 \text{ N} = 21.7 \text{ kN} \qquad\qquad Ans.$$

Equations of Motion: Rotation About a Fixed Axis

17-7. The 15-lb rod is pinned at its end and has an angular velocity of $\omega = 5$ rad/s when it is in the horizontal position shown. Determine the rod's angular acceleration and the pin reactions at this instant.

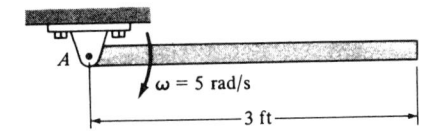

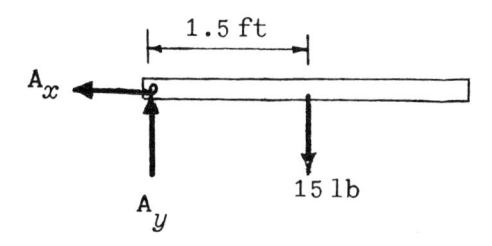

Solution

$\zeta + \Sigma M_A = I_A \alpha;$ _____

$$\alpha = 16.1 \text{ ft/s}^2 \qquad Ans.$$

$\xleftarrow{+} \Sigma F_x = m(a_G)_x;$ _____

$$A_x = 17.5 \text{ lb} \qquad Ans.$$

$+\uparrow \Sigma F_y = m(a_G)_y;$ _____

$$A_y = 3.75 \text{ lb} \qquad Ans.$$

17-8. A cord is wrapped around the inner core of a spool. If the cord is pulled with a constant tension of 30 lb and the spool is originally at rest, determine the spool's angular velocity when $s = 8$ ft of cord have unraveled. Neglect the weight of the cord. The spool and cord have a total weight of 400 lb and the radius of gyration about the axle A is $k_A = 1.30$ ft.

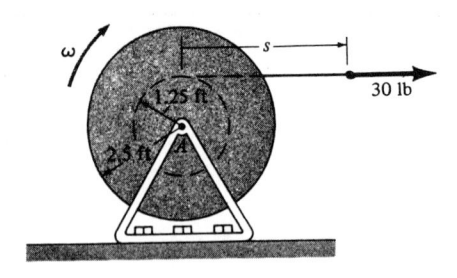

Solution

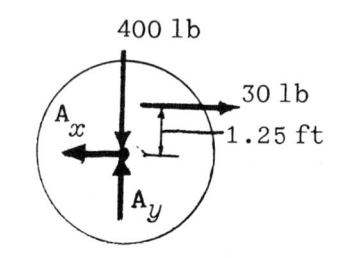

$\circlearrowleft + \Sigma M_A = I_A \alpha;$ _____

$$\alpha = 1.79 \text{ rad/s}^2$$

$$\theta = \frac{8}{1.25} = 6.40 \text{ rad}$$

$\circlearrowleft + \omega^2 = \omega_0^2 + 2\alpha_C(\theta - \theta_0)$

$\omega^2 = 0 + 2(1.79)(6.40 - 0)$

$\omega = 4.79$ rad/s *Ans.*

17-9. If the support at B is suddenly removed, determine the initial reactions at the pin A. The plate has a weight of 30 lb.

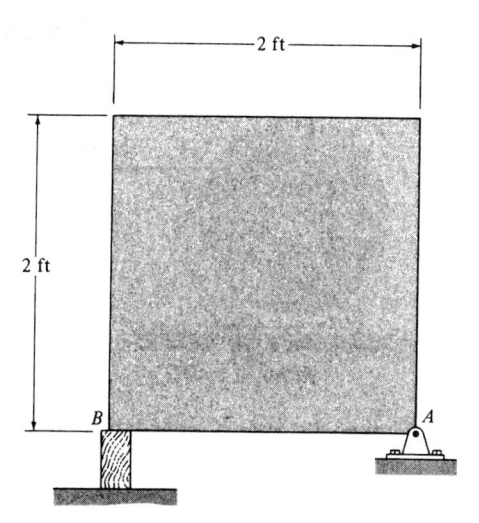

Solution

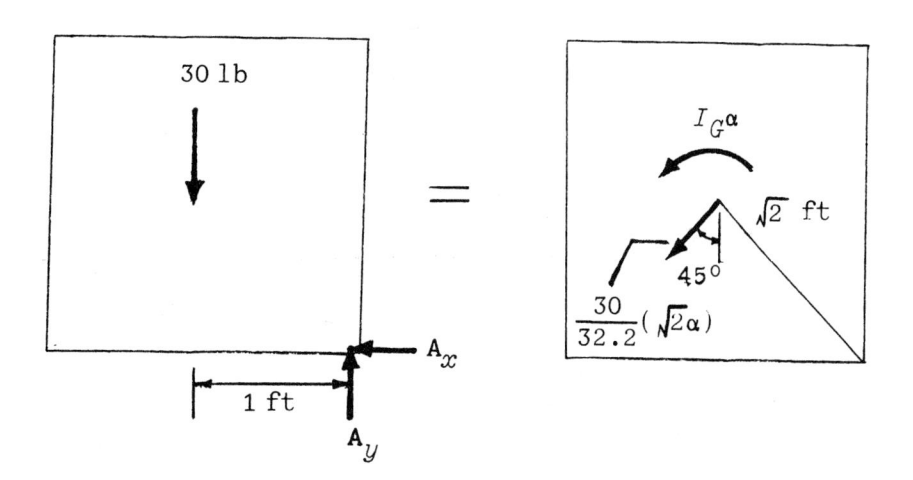

$$\xleftarrow{+} \Sigma F_x = m(a_G)_x; \quad \underline{\hspace{4cm}}$$

$$+\uparrow \Sigma F_y = m(a_G)_y; \quad \underline{\hspace{4cm}}$$

$$\left(+ \Sigma M_A = \Sigma (M_k)_A; \quad \underline{\hspace{4cm}}\right.$$

Solving,

$\alpha = 12.1 \text{ rad/s}^2$

$A_x = 11.25 \text{ lb}$ \hfill Ans.

$A_y = 18.75 \text{ lb}$ \hfill Ans.

17-10. The disk has a mass of 20 kg and is originally spinning at the end of the strut with an angular velocity of $\omega = 60$ rad/s. If it is then placed against the wall, for which $\mu_A = 0.3$, determine the time required for the motion to stop. What is the force in strut BC during this time?

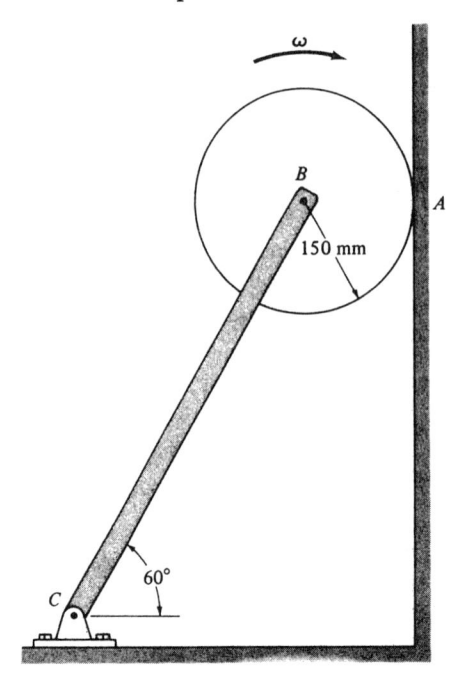

Solution

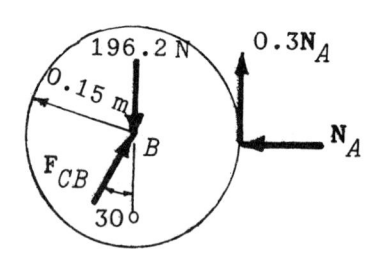

$(+\circlearrowleft \Sigma M_B = I_B \alpha;$ _____

$\overset{+}{\rightarrow} \Sigma F_x = m(a_B)_x;$ _____

$+\uparrow \Sigma F_y = m(a_B)_y;$ _____

Solving

$N_A = 96.6$ N, $\alpha = 19.3$ rad/s² $\circlearrowright$

$F_{CB} = 193$ N Ans.

$(+\circlearrowleft \omega = \omega_0 + \alpha t$

$0 = 60 - 19.3t$

$t = 3.11$ s Ans.

17-11. A clown, mounted on stilts, loses his balance and falls backward from the vertical position, where it is assumed that $\omega = 0$ when $\theta = 0°$. Paralyzed with fear, he remains *rigid* as he falls. His mass including the stilts is 80 kg, the mass center is at G, and the radius of gyration about G is $k_G = 1.2$ m. Determine the coefficient of friction between his shoes and the ground at A if it is observed that slipping occurs when $\theta = 30°$.

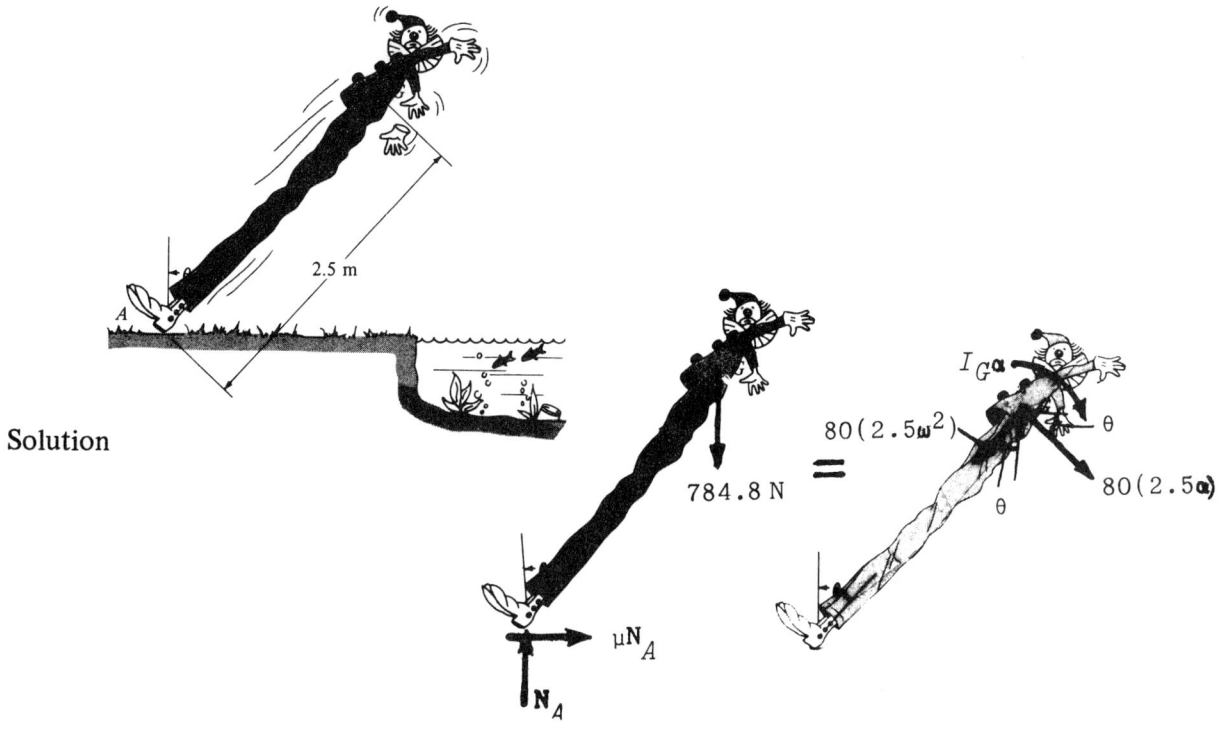

Solution

$\xrightarrow{+} \Sigma F_x = m(a_G)_x;$ _____ (1)

$+\uparrow \Sigma F_y = m(a_G)_y;$ _____ (2)

$(+ \Sigma M_A = \Sigma(M_k)_A;$ _____

Thus, $\qquad \alpha = 3.19 \sin\theta$

Kinematics,

$$\alpha d\theta = \omega d\omega$$

$$\int_0^\theta 3.19 \sin\theta \, d\theta = \int_0^\omega \omega \, d\omega$$

$$3.19[-\cos\theta]_0^\theta = \frac{1}{2}\omega^2$$

$$6.38[1 - \cos\theta] = \omega^2$$

At $\theta = 30°$, $\alpha = 1.595$ rad/s², $\omega = 0.925$ rad/s

From Eqs. (1) and (2),

$$N_A = 477 \text{ N}$$

$$\mu_A = 0.400 \qquad\qquad Ans.$$

Equations of Motion: General Plane Motion

17-12. The wheel has a weight of 30 lb, a radius of $r = 0.5$ ft, and a radius of gyration of $k_G = 0.23$ ft. If the coefficient of friction between the wheel and the plane is $\mu = 0.2$, determine the wheel's angular acceleration as it rolls down the incline. Set $\theta = 12°$.

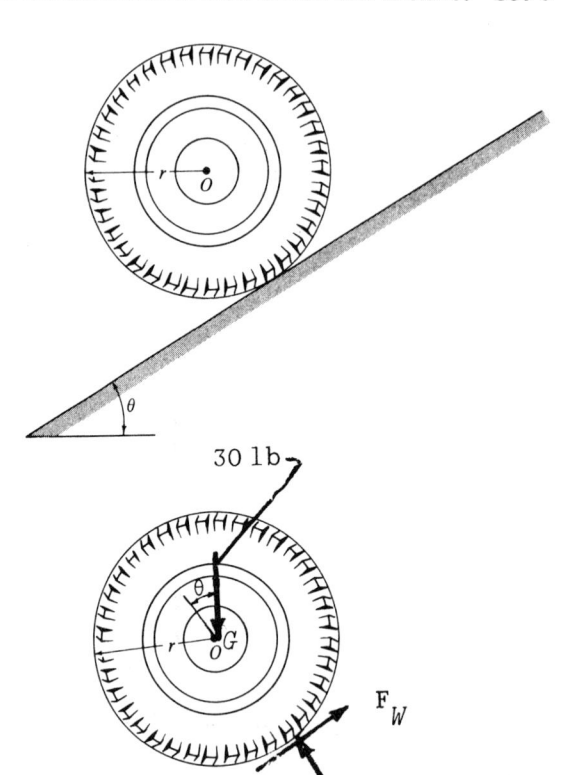

Solution

$+\swarrow \Sigma F_x = m(a_G)_x;$ _____ (1)

$+\nwarrow \Sigma F_y = m(a_G)_y;$ _____ (2)

$(+\circlearrowleft \Sigma M_G = I_G \alpha;$ _____ (3)

Assume the wheel does not slip

$$\therefore a_G = (0.5)\alpha \qquad (4)$$

Solving Eqs. (1)–(4) yields

$$F_W = 1.0895 \text{ lb} \quad N_W = 29.34 \text{ lb} \quad a_G = 5.525 \text{ ft/s}^2$$

$$\alpha = 11.05 \text{ rad/s}^2 \qquad \qquad Ans.$$

Since $F_{W_{max}} = (0.2)(29.34) = 5.87 \text{ lb} > F_W$ Eq. (4) is valid

17-13. The slender 200-kg beam is suspended by a cable at its end as shown. If a man pushes on its other end with a horizontal force of 30 N, determine the initial acceleration of its mass center G, the beam's angular acceleration, and the tension in the cable AB.

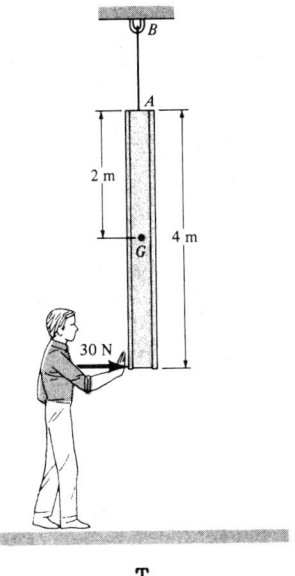

Solution

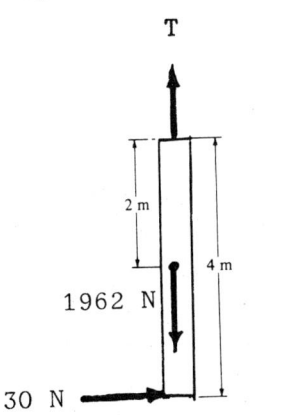

$\xrightarrow{+} \Sigma F_x = m(a_G)_x;$ _____

$+\uparrow \Sigma F_y = m(a_G)_y;$ _____

$\zeta+ \Sigma M_G = I_G \alpha;$ _____

Solving;

$T = 1962$ N Ans.

$a_G = 0.150$ m/s² Ans.

$\alpha = 0.225$ rad/s² Ans.

128 Study Guide and Problems

17-14. A woman sits in a rigid position on her rocking chair by keeping her feet on the bottom rungs at B. At the instant shown, she has reached an extreme backward position and has zero angular velocity. Determine her forward angular acceleration α and the frictional force at A necessary to prevent the rocker from slipping. The woman and the rocker have a combined weight of 180 lb and a radius of gyration about G of $k_G = 2.2$ ft.

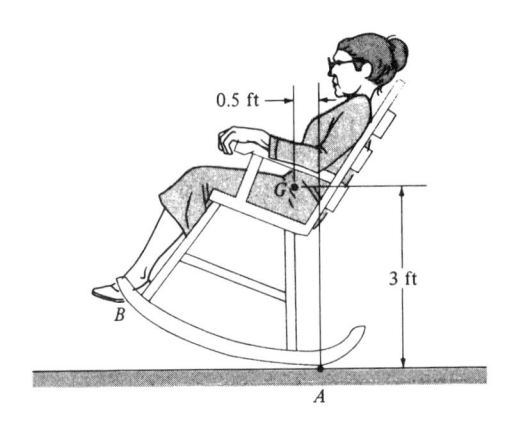

Solution

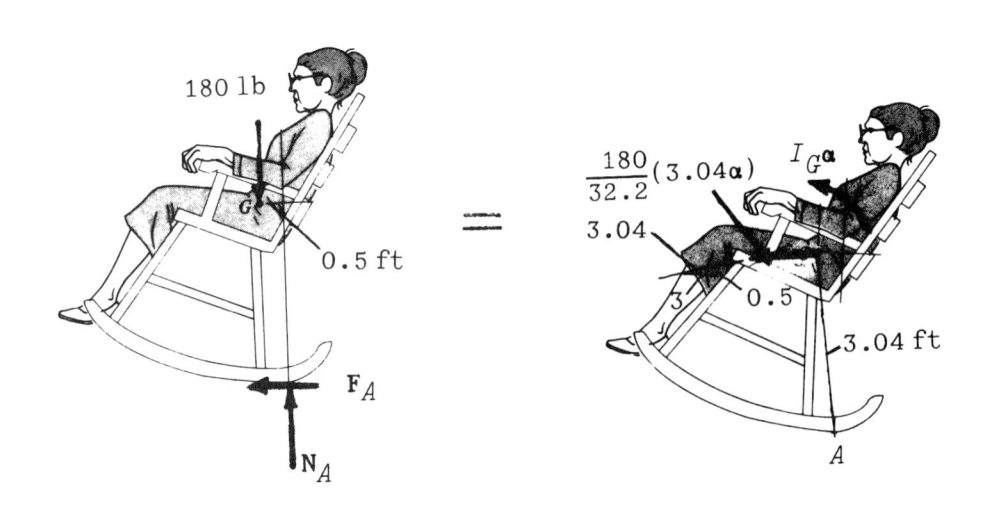

$\xrightarrow{+} \Sigma F_x = m(a_G)_x;$ _____

$+\uparrow \Sigma F_y = m(a_G)_y;$ _____

$\left(+ \Sigma M_A = \Sigma (M_k)_A;\right.$ _____

Solving;

$\alpha = 1.14$ rad/s²   *Ans.*

$F_A = 19.2$ lb   *Ans.*

$N_A = 177$ lb

17-15. A spool and the telephone wire wrapped around its core have a total weight of 80 lb and a radius of gyration of $k_G = 0.75$ ft. If the coefficient of friction between the spool and the ground is $\mu_A = 0.4$, determine the angular acceleration of the spool if the end of the cable is subjected to a horizontal force of 30 lb.

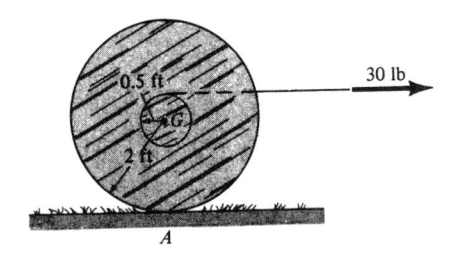

Solution

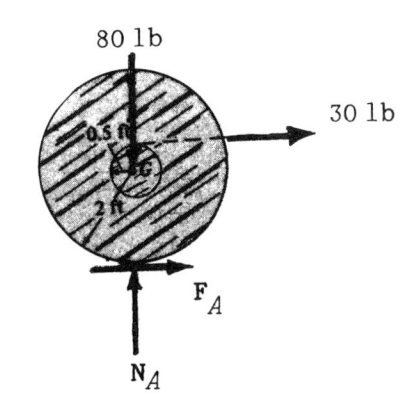

$\xrightarrow{+} \Sigma F_x = m(a_G)_x;$ _____

$+\uparrow \Sigma F_y = m(a_G)_y;$ _____

$\zeta + \Sigma M_G = I_G \alpha;$ _____

Assume no slipping occurs:

$$a_G = 2\alpha$$

Then,

$$N_A = 80 \text{ lb}$$
$$\alpha = 6.62 \text{ rad/s}^2 \quad \textit{Ans.}$$
$$F_A = 2.88 \text{ lb}$$

Since

$$(F_A)_{max} = (0.4)(80)$$
$$(F_A)_{max} = 32 \text{ lb} > 2.88$$

No slipping as assumed.

17-16. If the cable CB is horizontal and the beam is at rest in the position shown, determine the tension in the cable at the instant the towing force $F = 1500$ N is applied. The coefficient of friction between the beam and the floor at A is $\mu_A = 0.3$. For the calculation, assume that the beam is a uniform slender rod having a mass of 100 kg.

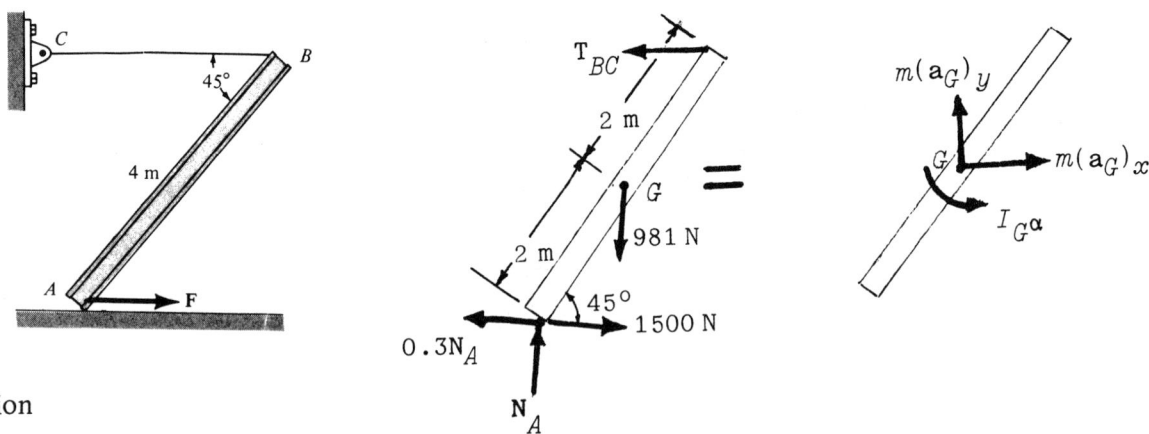

Solution

$\xrightarrow{+} \Sigma F_x = m(a_G)_x;$ _____ (1)

$(+\Sigma M_G = I_G \alpha;$ _____ (2)

$+\uparrow \Sigma F_y = m(a_G)_y;$ _____ (3)

Kinematics, $\quad \omega = 0$

$$\mathbf{a}_B = \mathbf{a}_A + \boldsymbol{\alpha} \times \mathbf{r}_{B/A} - \omega^2 \mathbf{r}_{B/A}$$

$$a_B \mathbf{j} = a_A \mathbf{i} + (\alpha \mathbf{k}) \times (4\cos 45° \mathbf{i} + 4\sin 45° \mathbf{j})$$

$$0 = a_A - \alpha 4 \sin 45°$$

$$a_A = 2.828 \alpha$$

$$\mathbf{a}_G = \mathbf{a}_A + \boldsymbol{\alpha} \times \mathbf{r}_{G/A} - \omega^2 \mathbf{r}_{G/A}$$

$$(a_G)_x \mathbf{i} + (a_G)_y \mathbf{j} = 2.828\alpha \mathbf{i} + (\alpha \mathbf{k}) \times (2\cos 45° \mathbf{i} + 2\sin 45° \mathbf{j})$$

$\xrightarrow{+} (a_G)_x = 2.828\alpha - 2\alpha \sin 45° = 1.414\alpha$

$+\uparrow (a_G)_y = 2\alpha \cos 45° = 1.414\alpha$

Substituting into Eqs. (1)–(3) yields

$$T_{BC} = 636 \text{ N}; \qquad\qquad\qquad\qquad\qquad\qquad\qquad\qquad\qquad Ans.$$

18 Planar Kinetics of a Rigid Body: Work and Energy

Principle of Work and Energy

18-1. An 800-lb tree falls from the vertical position such that it pivots about its cut section at A. If the tree can be considered as a uniform rod, pin-supported at A, determine the speed of its top branch B just before it strikes the ground.

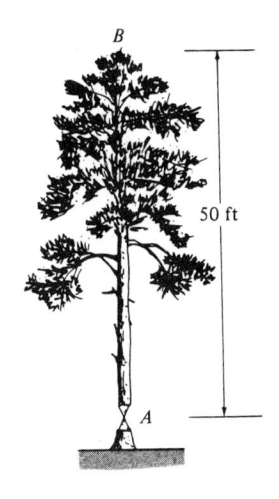

Solution

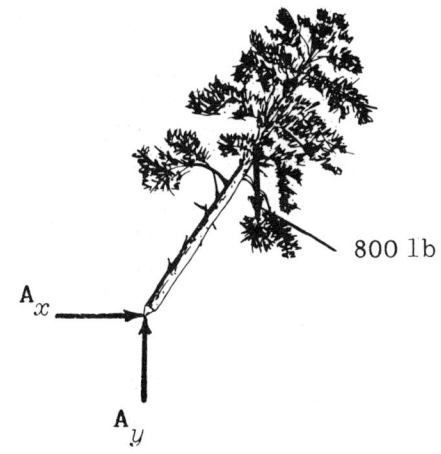

$T_1 + \Sigma U_{1-2} = T_2$

$\omega = 1.39$ rad/s

$v_B = 1.39(50)$

$v_B = 64.5$ ft/s *Ans.*

18-2. A motor supplies a constant torque or twist of $M = 120$ lb $\cdot$ ft to the drum. If the drum has a weight of 30 lb and a radius of gyration of $k_O = 0.8$ ft, determine the speed of the 15-lb crate A after it rises $s = 4$ ft starting from rest. Neglect the weight of the cord.

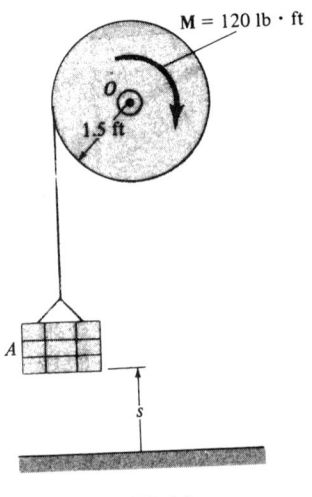

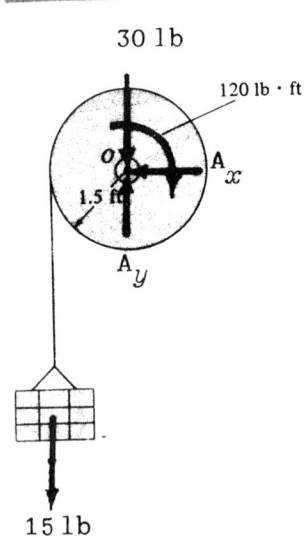

Solution

$T_1 + \Sigma U_{1-2} = T_2$

$v = 26.7$ ft/s Ans.

18-3. The spool of cable, originally at rest, has a mass of 200 kg and a radius of gyration of $k_G = 325$ mm. If the spool rests on two small rollers A and B and a constant horizontal force of $P = 400$ N is applied to the end of the cable, compute the angular velocity of the spool when 8 m of cable has been unraveled. Neglect friction and the mass of the rollers and unraveled cable.

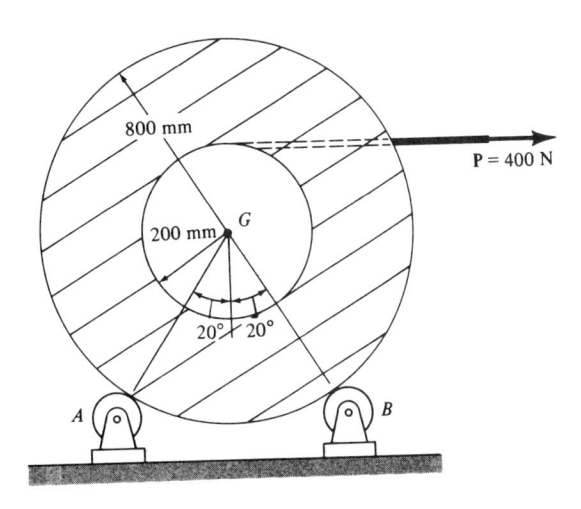

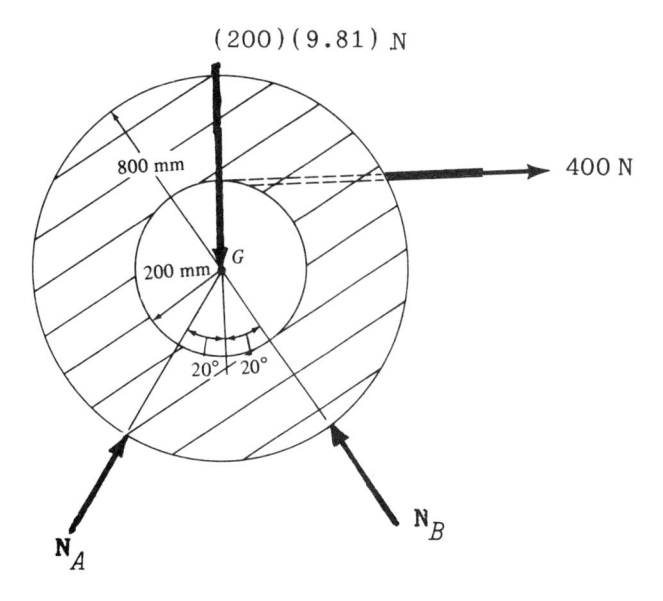

Solution

$T_1 + \Sigma U_{1-2} = T_2$

$\omega_2 = 17.4$ rad/s   *Ans.*

18-4. A man having a weight of 180 lb sits in a chair of the Ferris wheel, which has a weight of 15,000 lb and a radius of gyration of $k_O = 37$ ft. If a torque of $M = 80(10^3)$ lb·ft is applied about O, determine the angular velocity of the wheel after it has rotated 180°. Neglect the weight of the chairs and note that the man remains in an upright position as the wheel rotates. The wheel starts from rest in the position shown.

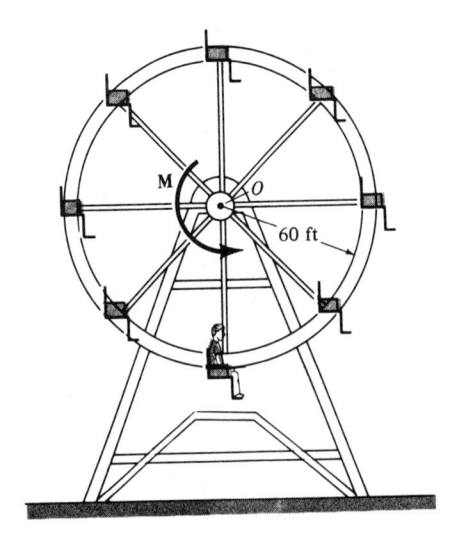

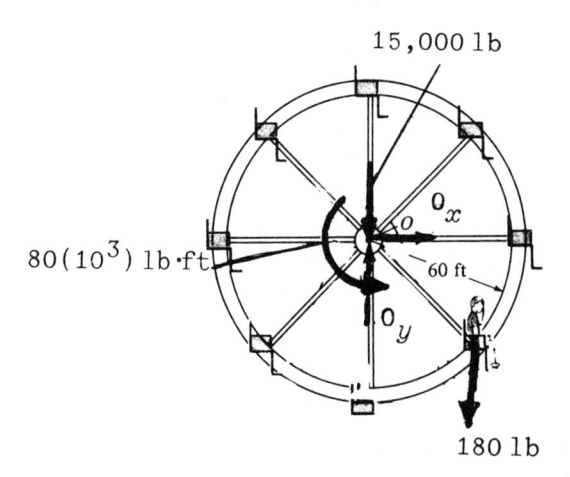

Solution

$$T_1 + \Sigma U_{1-2} = T_2$$

$\omega = 0.836$ rad/s *Ans.*

18-5. The beam having a weight of 150 lb is supported by two cables. If the cable at end B is cut so that the beam is released from rest when $\theta = 30°$, determine the speed at which end A strikes the wall. Neglect friction at B. Consider the beam to be a thin rod.

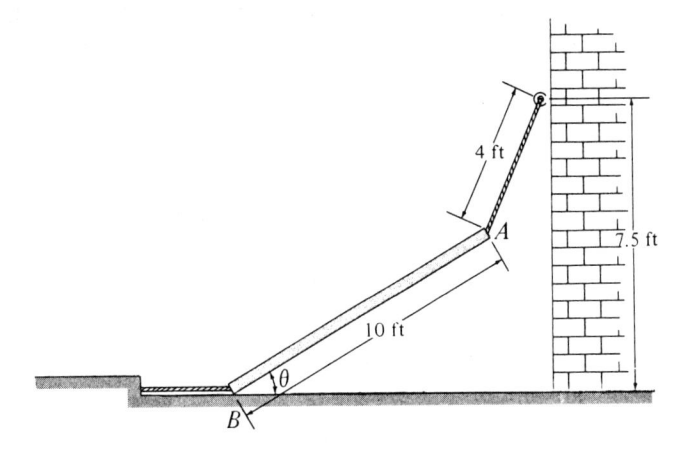

Solution

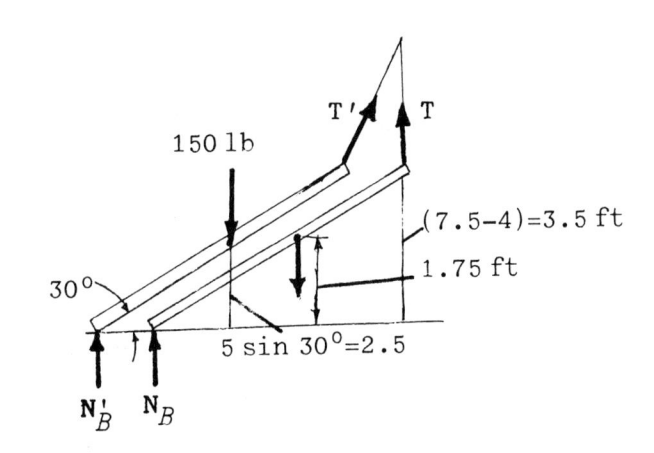

In the final position the *IC* is at infinity, $\omega = 0$.

$$T_1 + \Sigma U_{1-2} = T_2$$

$$v_G = 6.95 \text{ ft/s}$$

Thus,

$$v_A = 6.95 \text{ ft/s} \rightarrow \qquad \qquad Ans.$$

136 Study Guide and Problems

Conservation of Energy

18-6. If the 3-lb solid ball is released from rest when $\theta = 30°$, determine its angular velocity when $\theta = 0°$, which is the lowest point of the curved path having a radius of 11.5 in. The ball does not slip as it rolls.

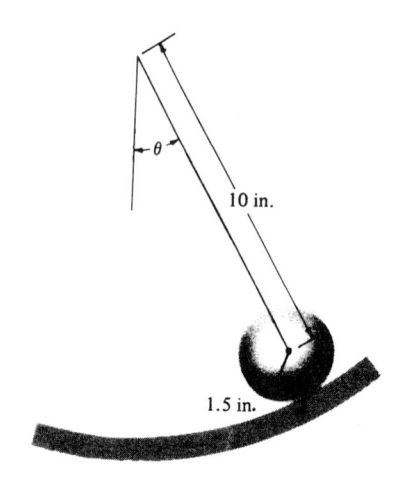

Solution

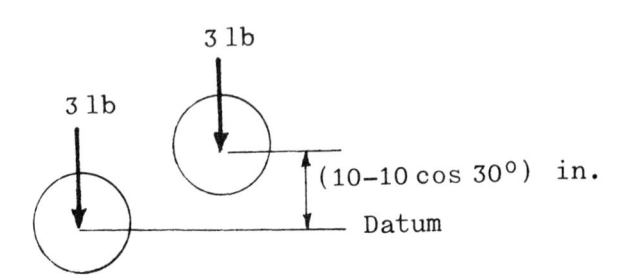

$$T_1 + V_1 = T_2 + V_2$$

$$v_G = \left(\frac{1.5}{12}\right) \omega$$

$$\omega = 18.1 \text{ rad/s} \qquad \qquad Ans.$$

18-7. The 500-g rod AB rests along the smooth inner surface of a hemispherical bowl. If the rod is released from rest from the position shown, determine its angular velocity ω at the instant it swings downward and becomes horizontal.

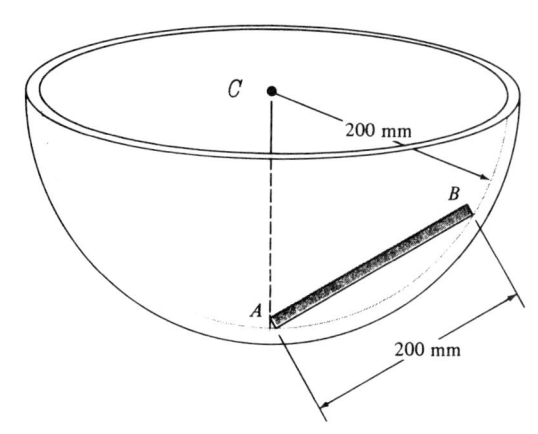

Solution

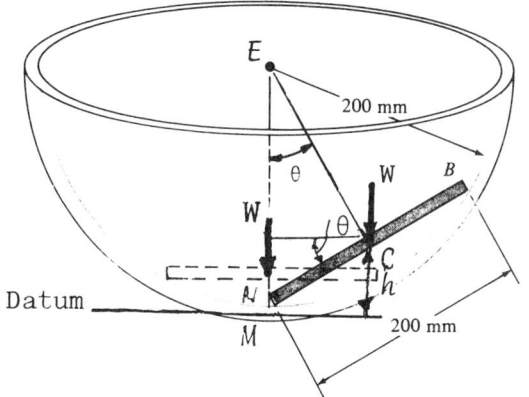

Select Datum through the bottom of the bowl.

$$\theta = \sin^{-1}\left(\frac{0.1}{0.2}\right) = 30°$$

$$h = 0.1 \sin 30° = 0.05$$

$$CE = \sqrt{(0.2)^2 - (0.1)^2} = 0.173$$

$$MN = 0.2 - 0.173 = 0.0268$$

$$T_1 + V_1 = T_2 + V_2$$

Since

$$v_G = 0.173\,\omega_{AB}$$

$$0.24525 = 0.131454 + 0.008316\,\omega_{AB}^2$$

$$\omega_{AB} = 3.70 \text{ rad/s} \hspace{2cm} Ans.$$

18-8. The small bridge consists of an 1,800-lb uniform deck *EF* (thin plate), two overhead beams *AB* (slender rods), each having a weight of 200 lb, and a 2,400-lb counterweight *BC*, which can be considered as a thin plate having the dimensions shown. The weight of the tie rods *AE* can be neglected. If the operator lets go of the rope when the bridge is at an at-rest position, $\theta = 45°$, determine the speed at which the end of the deck *E* hits the roadway step at *H*, $\theta = 0°$. The bridge is pin-connected at *A*, *D*, *E*, and *F*.

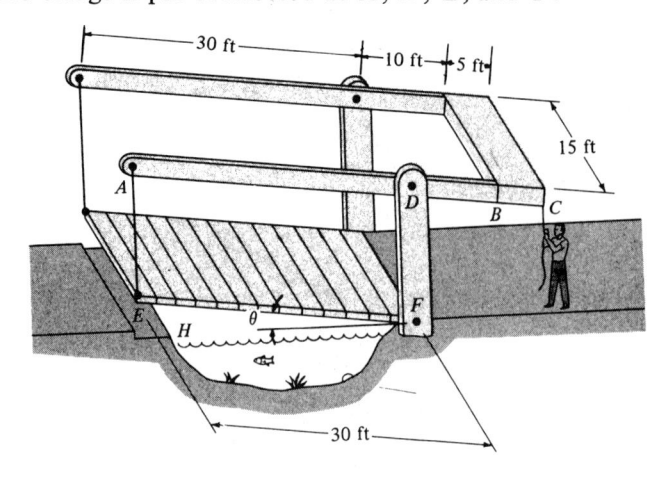

Solution

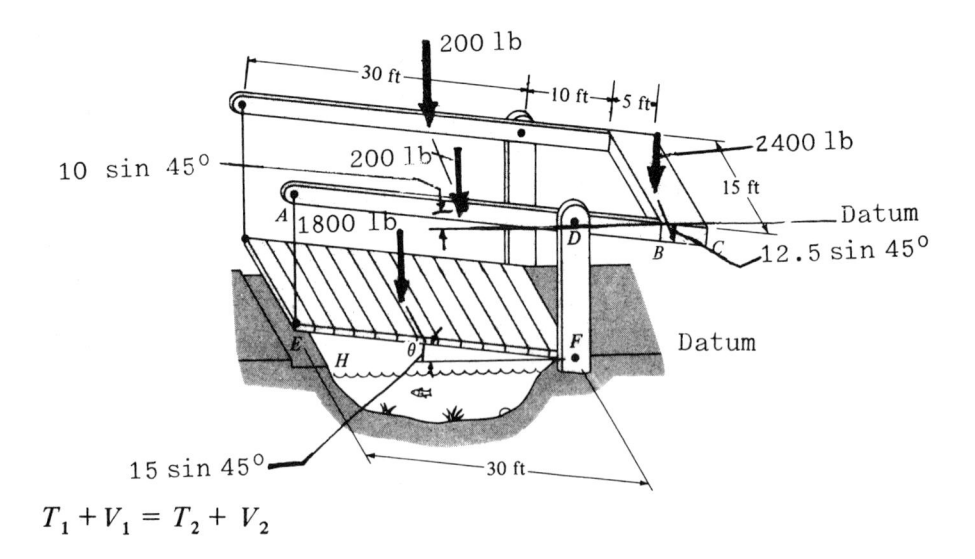

$$T_1 + V_1 = T_2 + V_2$$

Solving,

$$\omega = 0.212 \text{ rad/s}$$
$$v_E = \omega(30) = 0.212(30)$$
$$V_E = 6.36 \text{ ft/s} \qquad \text{Ans.}$$

18-9. A chain that has a negligible mass is draped over a sprocket which has a mass of 2 kg and a radius of gyration of $k_O = 50$ mm. If the 4-kg block A is released from rest in the position shown, $s = 1$ m, determine the angular velocity which the chain imparts to the sprocket when $s = 2$ m.

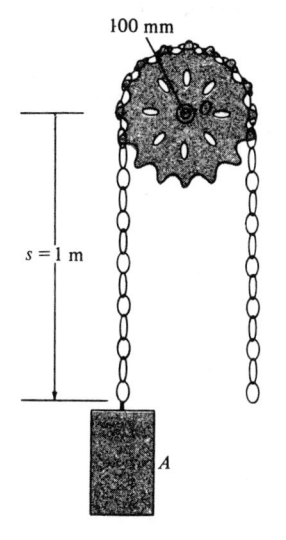

Solution

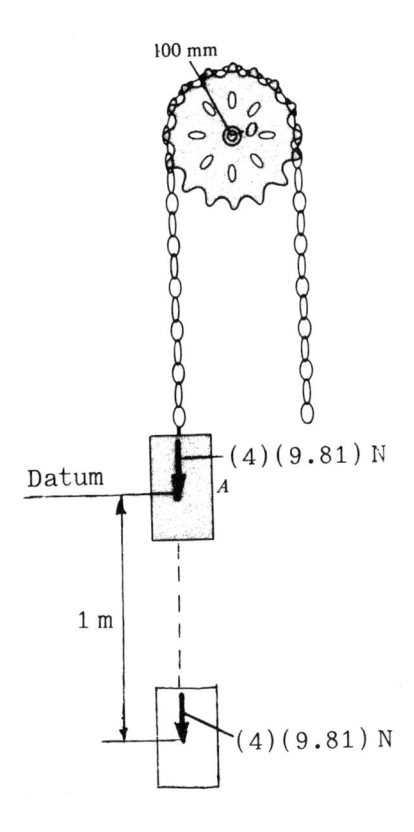

$$T_1 + V_1 = T_2 + V_2$$

Since $v_A = 0.1\,\omega$, substituting and solving for ω yields:

$$\omega = 41.8 \text{ rad/s} \qquad Ans.$$

140 Study Guide and Problems

18-10. The uniform slender rod has a mass of 5 kg. Determine the reaction at the pin O when the cord at A is cut and $\theta = 90°$.

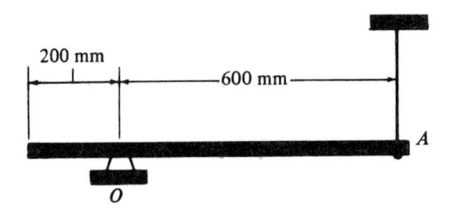

Solution

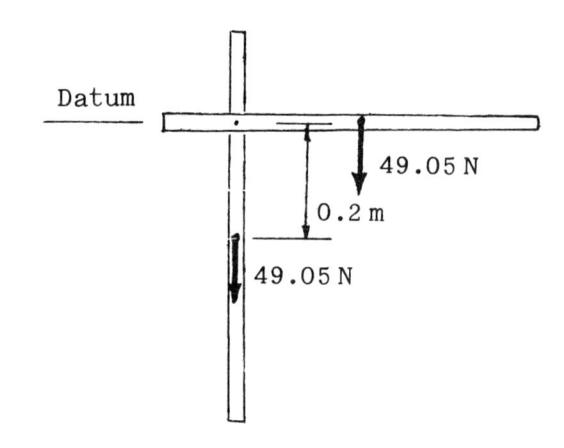

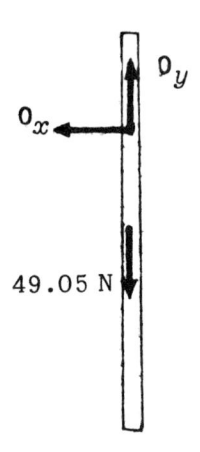

$$T_1 + V_1 = T_2 + V_2$$

$\omega = 6.48$ rad/s

$\xrightarrow{+} \Sigma F_x = m(a_G)_x;$ _____

$+\uparrow \Sigma F_y = m(a_G)_y;$ _____

$\zeta + \Sigma M_O = I_O \alpha;$ _____ , $\alpha = 0$

$O_x = 0 \quad O_y = 91.1$ N   Ans.

19 Planar Kinetics of a Rigid Body: Impulse and Momentum

Principle of Impulse and Momentum

19-1. A cord of negligible mass is wrapped around the outer surface of the 50-lb cylinder and its end is subjected to a constant horizontal force of $P = 2$ lb. If the cylinder rolls without slipping at A, determine its angular velocity in 4 s starting from rest. Neglect the thickness of the cord.

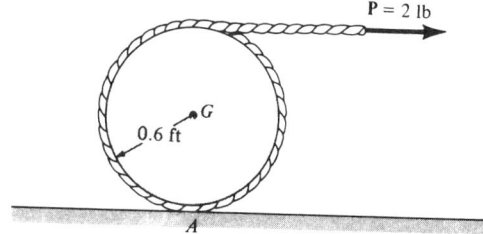

Solution

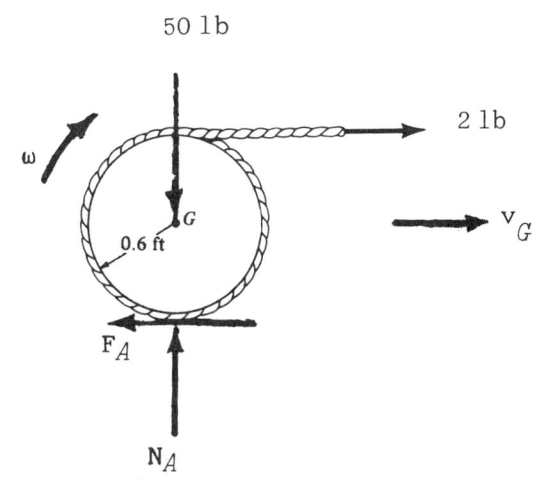

$(+\ (H_G)_1 + \Sigma \int M_G\, dt = (H_G)_2$

$\stackrel{+}{\rightarrow} m(v_G)_1 + \Sigma \int F_x\, dt = m(v_G)_2$

Solving

$v_G = 0.6\, \omega$

$\omega = 11.4$ rad/s *Ans.*

$v_G = 6.87$ ft/s

19-2. A cord of negligible mass is wrapped around the outer surface of the 2-kg disk. If the disk is released from rest, determine its angular velocity in 3 s.

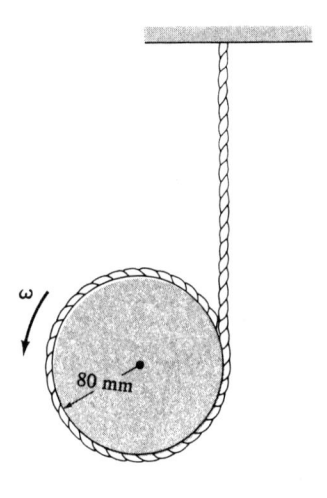

Solution

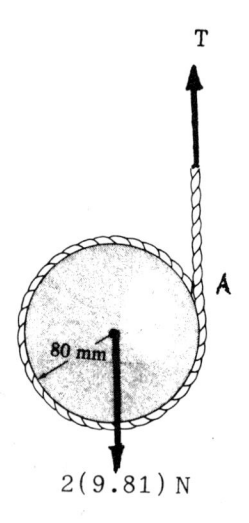

$$(\curvearrowleft+ \ (H_A)_1 + \Sigma \int M_A \, dt = (H_A)_2$$

$v_G = 0.08 \omega$ $\omega = 245$ rad/s   *Ans.*

19-3. A constant torque or twist of $M = 0.4$ N·m is applied to the center gear A. If the system starts from rest, determine the angular velocity of each of the three (equal) smaller gears in 3 s. The smaller gears (B) are pinned at their centers, and the mass and centroidal radii of gyration of the gears are given in the figure.

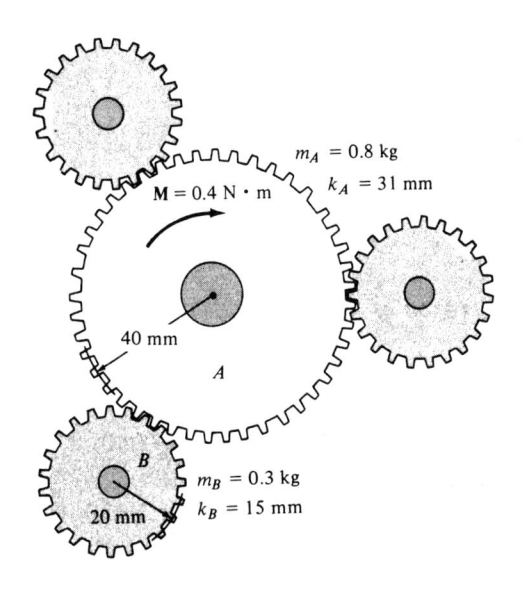

$m_A = 0.8$ kg
$k_A = 31$ mm
$M = 0.4$ N·m

$m_B = 0.3$ kg
$k_B = 15$ mm

Solution

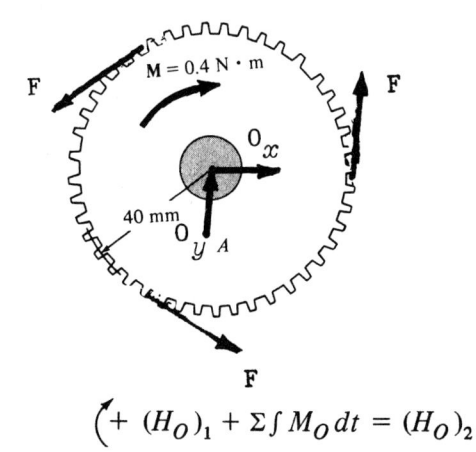

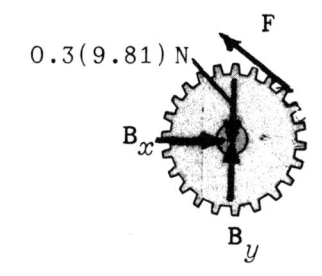

Gear A:

$$\left(\stackrel{+}{\curvearrowleft}\right) \ (H_O)_1 + \Sigma \int M_O \, dt = (H_O)_2$$

Gear B:

$$\left(\stackrel{+}{\curvearrowleft}\right) \ (H_B)_1 + \Sigma \int M_B \, dt = (H_B)_2$$

Since

$$0.04\omega_A = 0.02\omega_B \quad \text{or} \quad \omega_B = 2\omega_A$$

then solving,

$$\omega_A = 760 \text{ rad/s}; \quad F = 1.71 \text{ N}$$
$$\omega_B = 1520 \text{ rad/s} \qquad \qquad \textit{Ans.}$$

19-4. Gear A has a weight of 1.5 lb, a radius of 0.2 ft, and a radius of gyration of $k_O = 0.13$ ft. The coefficient of friction between the gear rack B and the horizontal surface is $\mu = 0.3$. If the rack has a weight of 0.8 lb and is initially sliding to the left with a velocity of $(v_B)_1 = 4$ ft/s, determine the constant moment **M** which must be applied to the gear to increase the motion of the rack so that in $t = 2.5$ s it will have a velocity of $(v_B)_2 = 8$ ft/s to the left. Neglect friction between the rack and the gear and assume that the gear exerts *only* a horizontal force on the rack.

Solution

$$m_A = \frac{1.5}{32.2} = 0.0466 \text{ slugs}; \quad m_B = \frac{0.8}{32.2} = 0.0248 \text{ slugs}$$

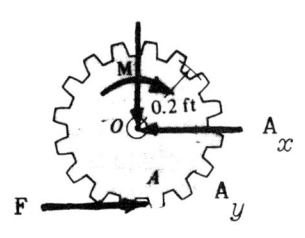

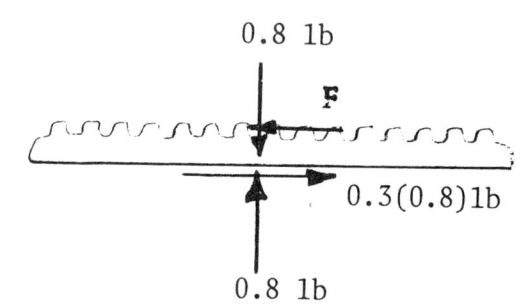

When

$$(v_B) = 4 \text{ ft/s}, \quad (\omega_A)_1 = \frac{4}{0.2} = 20 \text{ rad/s}$$

$$(v_B)_2 = 8 \text{ ft/s}, \quad (\omega_A)_2 = \frac{8}{0.2} = 40 \text{ rad/s}$$

$$(\stackrel{\curvearrowleft}{+}) \ (H_O)_1 + \Sigma \int M_O \, dt = (H_O)_2$$

$$\stackrel{+}{\leftarrow} \ m(v_x)_1 + \Sigma \int F_x \, dt = m(v_x)_2$$

$F = 0.280$ lb *Ans.*

$M = 0.0622$ lb · ft *Ans.*

19-5. The 12-kg disk has an angular velocity of $\omega = 20$ rad/s. If the brake ABC is applied such that the magnitude of force **P** varies with time as shown, determine the time needed to stop the disk. The coefficient of friction at B is $\mu = 0.4$.

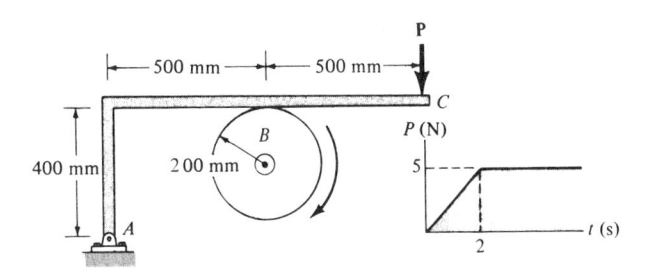

Solution

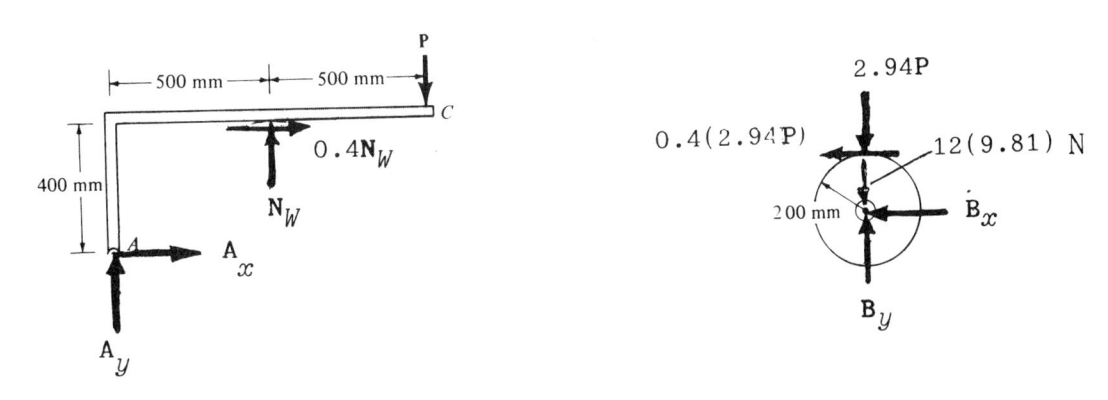

Brake:

$$\zeta + \Sigma M_A = 0;$$

$$N_W = 2.94 \, P$$

Disk:

$$\zeta + (H_B)_1 + \Sigma \int M_B \, dt = (H_B)_2$$

$$\left[\frac{1}{2}(12)(0.2)^2\right](20) - (0.2)\int_0^t 0.4(2.94P)\,dt = 0$$

$$\int_0^t P\,dt = 20.41$$

The integral is evaluated by finding an equivalent area under the curve. Thus, assuming $t > 2$ s

$$\int_0^t P\,dt = \frac{1}{2}(5)(2) + 5(t-2) = 20.41$$

Thus,

$$t = 5.08 \text{ s} \qquad\qquad Ans.$$

19-6. The flywheel A has a mass of 30 kg and a radius of gyration of $k_C = 95$ mm. Disk B has a mass of 25 kg, is pinned at D, and is coupled to the flywheel using a belt which is subjected to a tension such that it does not slip at its contacting surfaces. If a motor supplies a counterclockwise torque or twist to the flywheel, having a magnitude of $M = (12t)$ N·m, where t is measured in seconds, determine the angular velocity of the disk 3 s after the motor is turned on. Initially, the flywheel is at rest.

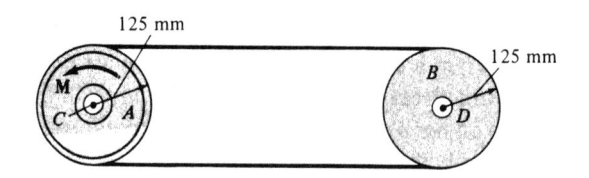

Solution

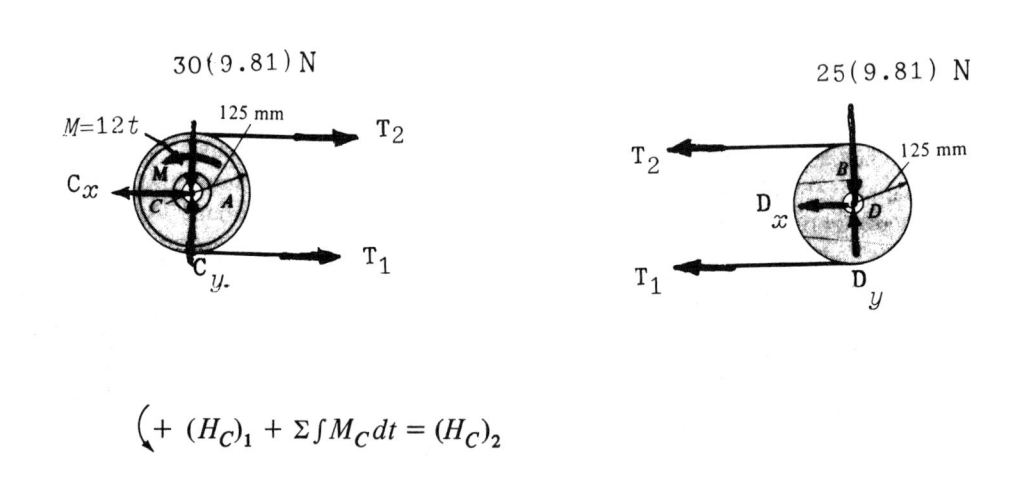

$$(\stackrel{+}{\curvearrowleft}) \quad (H_C)_1 + \Sigma \int M_C \, dt = (H_C)_2$$

$$(\stackrel{+}{\curvearrowleft}) \quad (H_D)_1 + \Sigma \int M_D \, dt = (H_D)_2$$

Note $\quad (\omega_A)_2 = (\omega_D)_2$

Solving,

$\quad (\omega_D)_2 = 116$ rad/s $\hfill$ *Ans.*

19-7. The 50-kg cylinder has an angular velocity of 30 rad/s when it is brought into contact with the horizontal surface at C. If the coefficient of friction is $\mu_C = 0.2$, determine how long it takes for the cylinder to stop spinning. What force is developed at the pin A during this time? The axis of the cylinder is connected to *two* symmetrical links. (Only AB is shown.) For the computation, neglect the weight of the links.

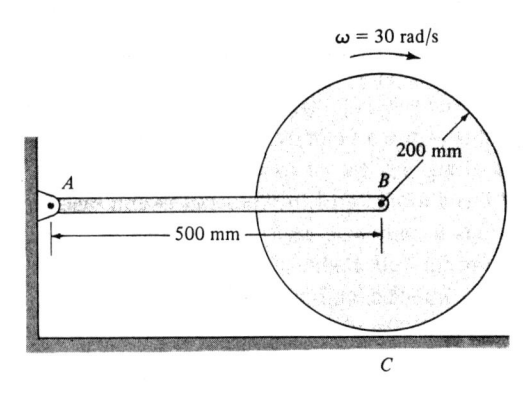

Solution

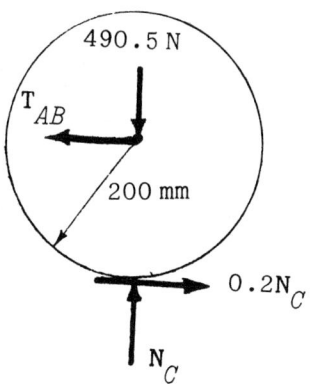

$$\xrightarrow{+} \quad m(v_G)_{x_1} + \Sigma \int F_x dt = m(v_G)_{x_2}$$

$$+\uparrow \quad m(v_G)_{y_1} + \Sigma \int F_y dt = m(v_G)_{y_2}$$

$$\zeta + (H_B)_1 + \Sigma \int M_B \, dt = (H_B)_2$$

Solving,

$t = 1.53$ s *Ans.*

$\dfrac{T_{AB}}{2} = 49.0$ N *Ans.*

$N_C = 490.5$ N

Conservation of Momentum

19-8. A horizontal circular platform has a weight of 300 lb and a radius of gyration about the z axis passing through its center O of $k_z = 8$ ft. The platform is free to rotate about the z axis and is initially at rest. A man, having a weight of 150 lb, begins to run along the edge in a circular path of radius 10 ft. If he has a speed of 4 ft/s and maintains this speed relative to the platform, compute the angular velocity of the platform.

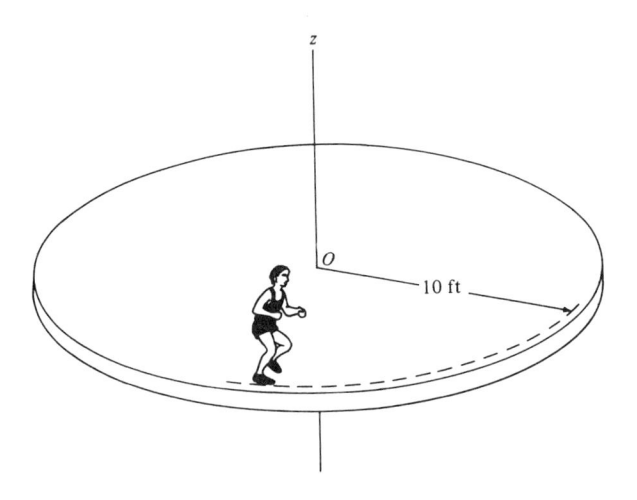

Solution

$$(H_z)_1 = (H_z)_2$$

$\omega = 0.175$ rad/s *Ans.*

19-9. The square plate, where $a = 0.75$ ft, has a weight of 4 lb and is rotating on the smooth surface with a constant angular velocity of $\omega_1 = 10$ rad/s. Determine the new angular velocity of the plate just after its corner strikes the peg P and the plate starts to rotate about P without rebounding.

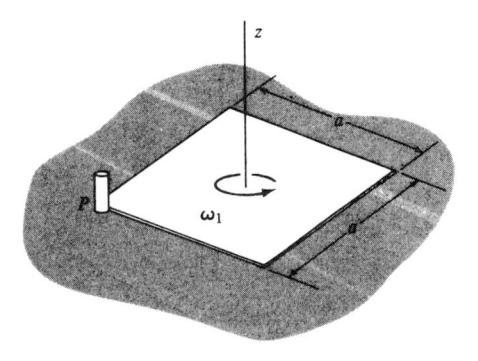

Solution

$$m = \frac{4}{32.2} = 0.1242 \text{ slugs}$$

$$I_G = \frac{1}{12} m[a^2 + a^2] = \frac{1}{6} ma^2$$

$$I_P = \frac{1}{6} ma^2 + m\left(\frac{a}{\sqrt{2}}\right)^2 = \frac{2}{3} ma^2$$

$$(H_P)_{z_1} = (H_P)_{z_2}$$

$$\omega_2 = 0.25 \omega_1$$

$$\omega_1 = 10 \text{ rad/s}$$

$$\therefore \omega_2 = 2.50 \text{ rad/s} \qquad \textit{Ans.}$$

19-10. The uniform pole has a mass of 15 kg and falls from rest when $\theta = 90°$ until it strikes the edge at A, $\theta = 60°$. If the pole then begins to pivot about this point after contact, determine the pole's angular velocity just after the impact. Assume that the pole does not slip at B as it falls until it strikes A.

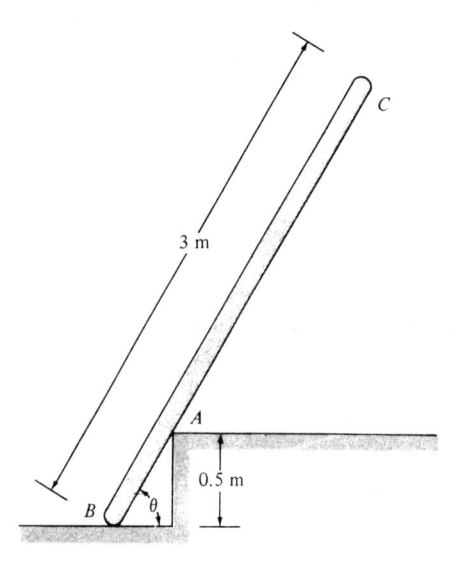

Solution

The datum is selected at B. Motion of the pole just before impact is

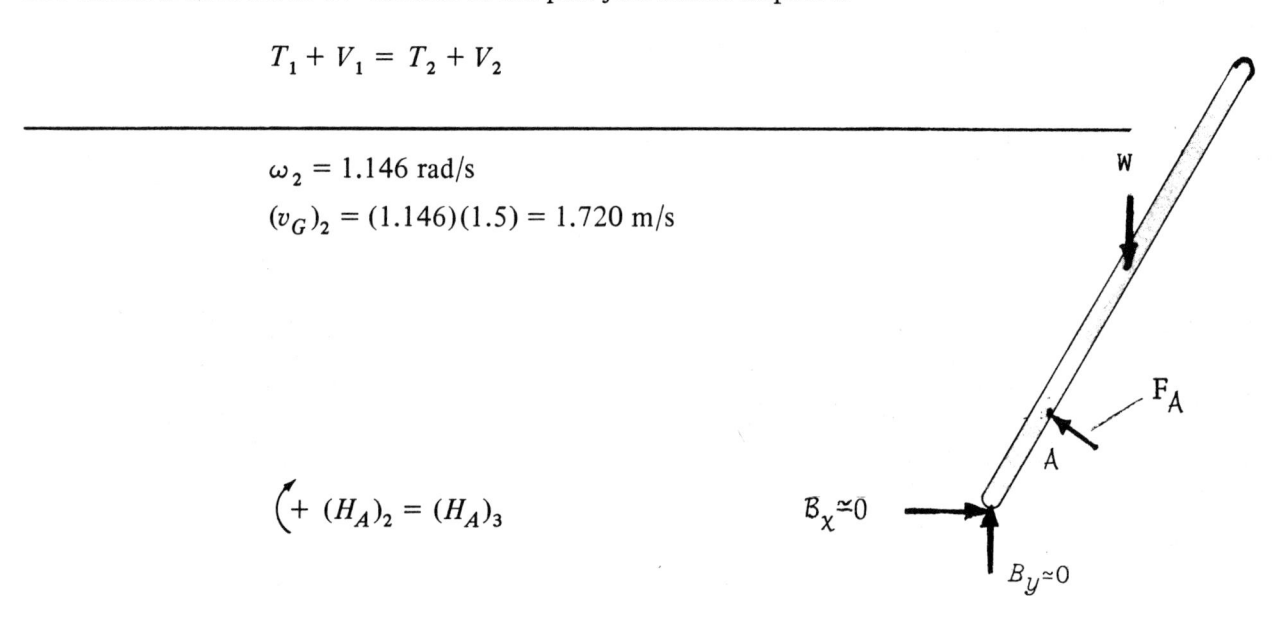

$$T_1 + V_1 = T_2 + V_2$$

$$\omega_2 = 1.146 \text{ rad/s}$$
$$(v_G)_2 = (1.146)(1.5) = 1.720 \text{ m/s}$$

$$(\stackrel{+}{\curvearrowleft}) \quad (H_A)_2 = (H_A)_3$$

Since

$$(v_G)_3 = \left[\frac{3}{2} - \frac{0.5}{\sin 60°}\right]\omega_3 = 0.9226\omega_3$$

$$\omega_3 = 1.53 \text{ rad/s} \qquad \qquad Ans.$$

19-11. The uniform rod AB has a weight of 3 lb and is released from rest without rotating from the position shown. As it falls, the end A strikes a hook S, which provides a permanent connection. Determine the speed at which the other end B strikes the wall at C.

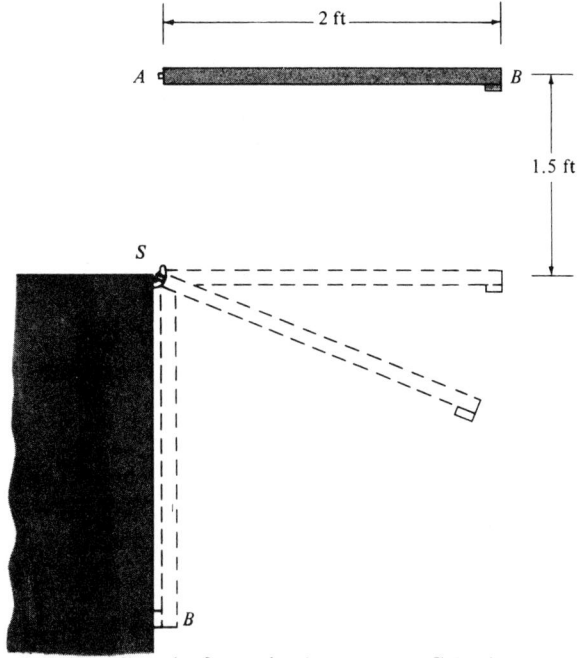

Solution

The speed of the rod's mass center just before the impact at S is determined from the conservation of energy. Putting the datum at the bar's initial position, we have

$$T_1 + V_1 = T_2 + V_2$$

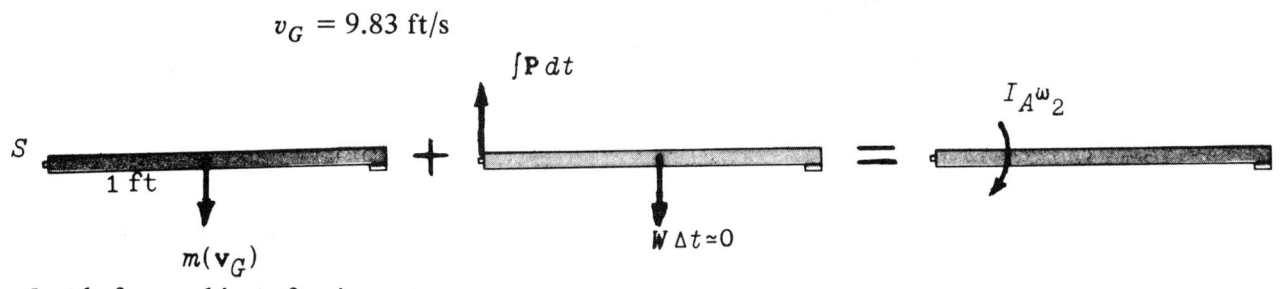

$$v_G = 9.83 \text{ ft/s}$$

Just before and just after impact

$$(\curvearrowleft +\ (H_S)_1 = (H_S)_2$$

$$\omega_2 = 7.37 \text{ rad/s}$$

Applying the conservation of energy theorem with the datum at the bar's horizontal position, we have

$$T_2 + V_2 = T_3 + V_3$$

$$\omega_3 = 10.13$$

Thus,

$$(v_B)_3 = (10.13)(2) = 20.2 \text{ ft/s} \qquad Ans.$$

Answers

12-1. $0 = 80 + (-10)t$

          $0 = (80)^2 + 2(-10)(s - 0)$

          $s = 0 + 80(8) + \frac{1}{2}(-10)(8)^2$

12-2. $\int_0^s ds = \int_0^t 2t\, dt$

12-3. $(0)^2 = (35)^2 + 2(-32.2)(h - 0)$

          $0 = 35 - 32.2t$

          $-60 = 0 + 35t + \frac{1}{2}(-32.2)t^2$

12-4. $948 - 114$

          $45t^2 - 3$

          $90t$

12-5. $(60)^2 = 0 + 2a_1(150)$

          $(100)^2 = (60)^2 + 2a_2(325 - 150)$

          $60 = 0 + 12t_1$

          $100 = 60 + (18.29)t_2$

          $\dfrac{(325 - 0)}{(5 + 2.19)}$

          $\dfrac{(100 - 0)}{(5 + 2.19)}$

12-6. $(10)(30) + \frac{1}{2}(10)(30)$

12-7. $\int_0^v dv = \int_0^t \frac{8}{10} t\, dt$

          $\int_{40}^v dv = \int_{10}^t 8\, dt$

154 Answers

12-8. $(18)(15)$

$(270) + (25)(20-15)$

$\frac{1}{2}(15)(270)$

$2025 + 270(20-15) + \frac{1}{2}(395-270)(20-15)$

12-9. $80/10$

$(80-80)/(40-10)$

$\frac{1}{2}(10)(80)$

$400 + 80(40-10)$

$\int_0^s ds = \int_0^t 8t\,dt$

$\int_{400}^s ds = \int_{10}^t 80\,dt$

12-10. $\frac{1}{3}s\left(\frac{1}{3}ds\right) = a\,ds$

$(200-s)(-ds) = a\,ds$

12-11. $1.25(40)^2$

$0.03(40)^3$

$2.5(40)$

$0.09(40)^2$

$\ddot{x} = 2.50$

$\ddot{y} = 0.18(t) = 0.18(40)$

12-12. 4

0

$-4x\dot{x}$

$-4\dot{x}^2 - 4x\ddot{x}$

12-13. $v_A \cos 30°$

$v_A \sin 30°$

$25 = v_A \cos 30° \, t$

$0 = 0 + v_A \sin 30° \, t + \frac{1}{2}(-9.81)t^2$

12-14. $18 = v_A \cos\theta \,(1.5)$
$0 = (v_A \sin\theta)^2 - 2(32.2)(h - 3.5)$
$0 = v_A \sin\theta - 32.2(1.5)$

12-15. $40 \sin 30°$
$40 \cos 30°$
$R \cos 30° = 20t$
$R \sin 30° = 0 - 34.64t + \dfrac{1}{2}(32.2)t^2$

12-16. $(5)^2/20$
2

12-17. $\left(\dfrac{45-30}{5}\right)\left(\dfrac{1000}{3600}\right)$
$\left(\dfrac{1}{200}\right)\left(\dfrac{40(1000)}{3600}\right)^2$

12-18. 4 m/s^2
$8^2/5.66$

12-19. $v = 0.15\,t^2$
$s = 0.05\,t^3$
$300(\pi/3)$
$(314.16/0.05)^{1/3}$
$0.15(18.45)^2$
$0.3(18.45)$
$(51.1)^2/300$

12-20. 32.2 ft/s^2
0
$(150)^2/32.2$
150 ft/s
$(v_B)_y^2 = 0 + 2(32.2)(1500 - 0)$

$32.2 \cos 64.23°$
$32.2 \sin 64.23°$
$(345)^2/14.0$

12-21. 0.2
0
0

12-22. 25
0
0
$0.3t$
0.3
0
$-8\cos(0.3t)$
$2.4\sin(0.3t)$
$0.72\cos(0.3t)$

12-23. $3t$
3
0
$500/\cos\theta$
$500\sec\theta\tan\theta\,\dot\theta$
$1500(\sec\theta\tan^2\theta + \sec^3\theta)\dot\theta$

12-24. 0.25
0
0
4
0
$0.25\cos\theta$
$-0.25\sin\theta\,\dot\theta$
$-\cos\theta\,\dot\theta$

12-25. $-4\sin 2\theta \dot{\theta}$
$-4\sin 2\theta \ddot{\theta} - 8\cos 2\theta \dot{\theta}^2$
0

12-26. $2s_B + s_A = l$
$-v_A/2$

12-27. $2s_C + (s_C - s_P) = l$
$3v_C = v_P$

12-28. $4(x_C) = 4(5) = 20$ m
$3x_C + \sqrt{x_D^2 + (5)^2}$
$3\dot{x}_C + \frac{1}{2}(x_D^2 + (5)^2)^{-1/2} 2x_D \dot{x}_D = 0$
$(v_D)^2 = 0 + 2(2)(3-0)$

12-29. $3/0.75$
$\underset{\rightarrow}{4} = 3\uparrow + v_{f/t}$

12-30. $-20\cos 45°\mathbf{i} + 20\sin 45°\mathbf{j} = 45\mathbf{i} + (v_{A/B})_x\mathbf{i} + (v_{A/B})_y\mathbf{j}$
$45 + (v_{A/B})_x$
$0 + (v_{A/B})_y$
$\frac{(20)^2}{0.2}\cos 45°\mathbf{i} + \frac{(20)^2}{0.2}\sin 45°\mathbf{j} = 1600\mathbf{i} + (a_{A/B})_x\mathbf{i} + (a_{A/B})_y\mathbf{j}$

12-31. $-18\mathbf{j} = 25\mathbf{i} + (v_{A/B})_x\mathbf{i} + (v_{A/B})_y\mathbf{j}$
$\frac{(18)^2}{300}\mathbf{i} + 1.5\mathbf{j} = 2\mathbf{i} + (a_{A/B})_x\mathbf{i} + (a_{A/B})\mathbf{j}$

12-32. $2s_H + s_P = l$
$2v_H = -v_P$
$12\downarrow = 6\uparrow + \mathbf{v}_{P/H}$

158 Answers

12-33. $0 + 120t + \frac{1}{2}(12)t^2$

$1500 + 120t + \frac{1}{2}(-3)t^2$

$120 + 12(14.14)$

$120 - 3(14.14)$

13-1. $7500 - 5500 = 72\,000\,a$

13-2. $2(9.81) - 20 = 2a$

13-3. $0.8t$

$T = 300(4)$

13-4. $N_B - 1.5\cos 30° = 0$

$-0.3(1.30) + 1.5\sin 30° = \frac{1.5}{32.2}a$

$v_2^2 = 0 + 2(7.728)(15)$

13-5. $5000 - 3000x - 2000x = 2a$

$(2500 - 2500x)dx$

13-6. $N_A - 80\cos 60° = 0$

$80\sin 60° - 0.2N_A - 2T = \frac{80}{32.2}a_A$

$-T + 20 = \frac{20}{32.2}a_B$

13-7. $T - 30 = \frac{30}{32.2}(6)$

$-A_x + 35.59 = 0$

$A_y - 200 - 35.59 = 0$

$-M_A + 200(2.5) + 35.59(5) = 0$

13-8. $15 = \dfrac{15}{32.2}\left(\dfrac{v^2}{4}\right)$

No, water will fall out tangent to the path of motion.

13-9. $N_T - 100(9.81)\cos 72.6° = 100\left(\dfrac{(4)^2}{94.2}\right)$

$100(9.81)\sin 72.6° = 100 a_t$

13-10. $5\cos\theta = \left(\dfrac{5}{32.2}\right) a_t$

$T - 5\sin\theta = \left(\dfrac{5}{32.2}\right)\left(\dfrac{v^2}{3}\right)$

13-11. $2(9.81)\sin\theta = 2a_t$

$-N_s + 2(9.81)\cos\theta = 2\left(\dfrac{v^2}{0.8}\right)$

13-12. $4 + 3t = 10$ m
3 m/s
0
$t^2 + 2 = 6$ rad
$2t = 4$ rad/s
2 rad/s²
$6 - t^3 = -2$ m
$-3t^2 = -12$ m/s
$-6t = -12$ m/s²

13-13. $N_C \cos 30° - 2(9.81)\sin 30° = 2a_r$
$N_C \sin 30° + F_C - 2(9.81)\cos 30° = 2a_\theta$
0
$-1.4 \sin\theta \dot\theta$
$-0.7 \cos\theta \dot\theta$

13-14. $2\theta/2$

$P \cos 57.52° - N_s \sin 57.52°$

$P \sin 57.52° + N_s \cos 57.52°$

4
0
$2(\pi/2) = \pi$
$2\dot\theta = 2(4) = 8$
$2\ddot\theta = 0$

13-15. $1.5 \sin\theta$

$\dfrac{1.5(2)}{1.5}$

$-N_C \cos 26.57°$

$N_C \sin 26.57° + F$

13-16. $F_T \cos\phi$

$F_T \sin\phi$

14-1. $\dfrac{1}{2}(2000)(8.33)^2 - 0.20 F_{avg} = 0$

14-2. $F_{avg} = 1500(29.43)$

$\dfrac{1}{2}(1500)(1.5)^2 - 44145(x) = 0$

14-3. $(s - 0.75)(60)$

14-4. $\dfrac{1}{2}\left(\dfrac{150}{32.2}\right)(4)^2 + 150(25 - 6) = \dfrac{1}{2}\left(\dfrac{150}{32.2}\right) v_B^2$

14-5. $\frac{1}{2}\left(\frac{2}{32.2}\right)(6)^2 + 15(0.75 + x) - 0.4(0.75 + x)$

$\quad -\frac{1}{2}(20)x^2 - \frac{1}{2}(40)(x - 0.25)^2 = 0$

14-6. $\frac{1}{2}(250)(3)^2 + 250(9.81)(16) = \frac{1}{2}(250)(v_B)^2$

$N_T - 250(9.81) = 250\left(\frac{(18.0)^2}{8}\right)$

14-7. $(50)(9.81)(6)$
$2943/3$
$981/4(10^3)$

14-8. $350(550) = 25\,000 \sin\theta\,(50)$

14-9. $60(9.81) + 3T - 400(9.81) = 0$
$2s_E + (s_E - s_P) = l$

14-10. $25(10^3)(9.81)(22.22)\sin 10°$

14-11. $0 + 1.5(10 - 0) = \frac{1}{2}\left(\frac{1.5}{32.2}\right)(v_B)^2 + 0$

14-12. $\frac{1}{2}(300)(0.1 + 0.05)^2 = \frac{1}{2}(0.25)(v_2)^2 + \frac{1}{2}(0.3)(v_2)^2 + \frac{1}{2}(300)(0.05)^2$

14-13. $0 + 2\left[\frac{1}{2}(150)\left(\sqrt{(2)^2 + (1.5)^2} - (0.5)\right)^2\right] =$

$\quad \frac{1}{2}\left(\frac{1.5}{32.2}\right)(v_2)^2 + 2\left[\frac{1}{2}(150)(2 - 0.5)^2\right]$

14-14. $\frac{1}{2}\left(\frac{350}{32.2}\right)(15)^2 + 0 = \frac{1}{2}\left(\frac{350}{32.2}\right)(v_B)^2 - 350(200)$

$$350\cos 63.43° - N_B = \frac{350}{32.2}\left(\frac{(114)^2}{1118.0}\right)$$

14-15. $\quad 0+0 = \frac{1}{2}\left(\frac{600}{32.2}\right)(v_C)^2 + \frac{1}{2}\left(\frac{200}{32.2}\right)\left(\frac{v_C}{2}\right)^2 + 200(15) - 600\sin 20°(30)$

15-1. $\quad -(0.2)(10) + \int F_x\,dt = (0.2)(20)\cos 40°$
$\quad\quad 0 + \int F_y\,dt = (0.2)(20)\sin 40°$

15-2. $\quad 0 + \int F_x\,dt = (0.04)(200)\sin 60°$
$\quad\quad 0 + \int F_y\,dt = (0.04)(200)\cos 60°$

15-3. $\quad 0 - T(2) + 2(2) = \frac{2}{32.2}(v_A)_2$
$\quad\quad 0 + 4(2) - T(2) = \frac{4}{32.2}(v_B)_2$
$\quad\quad 2s_A + 2s_B = l$

15-4. $\quad 0 + \int F\,dt - (0.5)(9.81)(75)(0.4) = 75(0.2)$

15-5. $\quad -\left(\frac{30}{32.2}\right)(6) + \int_0^{15} 25\cos\left(\frac{\pi}{10}t\right)dt = \frac{30}{32.2}(v_x)_2$

15-6. $\quad 0 + \int_{2.45}^{5} 60t\,dt + 300(6-5) - 147.2(6-2.45) = 15(v_y)_2$

15-7. $\quad 0 + 0 = (0.0015)(1400) - 2.5(v_R)_2$

15-8. $\quad 0.6(10)\cos 30° = (25.6)v$

15-9. $\quad 0 + 40(15) = \frac{1}{2}\left(\frac{40}{32.2}\right)(v_B)^2 + 0$

$\quad\quad 0 + \frac{40}{32.2}(31.08) = \left(\frac{40+20}{32.2}\right)v_2$

$$\frac{40}{32.2}(31.08) - \int F\,dt = 0$$

15-10. $\quad 0 + 0 = \left(\dfrac{-20}{32.2}\right) v_w + \left(\dfrac{90}{32.2}\right)(6\cos 30° - v_w)$

$0 + 0 + N_w(0.8) - 90(0.8) - 20(0.8) = 0 + \left(\dfrac{90}{32.2}\right)(6\sin 30°)$

15-11. $\quad 0 + 0 = -80 v_A + 70(2 - v_A)$

$70(2 - 0.933) = 150 v_B$

15-12. $\quad 0 + 0 = 8[2\cos 30° - (v_m)_2] - 70(v_m)_2$

$0 + 0 + R_{\text{Avg}}(1.5) - (70 + 8)(9.81)(1.5) = 8(2\sin 30°)$

15-13. $\quad 0 + 900(3) = \dfrac{1}{2}\left(\dfrac{900}{32.2}\right)(v_H)_1^2 + 0$

$\left(\dfrac{900}{32.2}\right)(13.90) + 0 = \left(\dfrac{900}{32.2}\right)(v_H)_2 + \dfrac{500}{32.2}(v_P)_2$

$0.6 = \dfrac{(v_P)_2 - (v_H)_2}{13.90}$

$\dfrac{1}{2}\left(\dfrac{500}{32.2}\right)(14.29)^2 + \dfrac{1}{2}(500)(1)^2 = 0 - 500 x_2 + \dfrac{1}{2}(500)(x_2 + 1)^2$

15-14. $\quad (0.75)(4) + 0 = -(0.75)(v_B)_2 + 2.0(v_A)_2$

$\dfrac{(v_A)_2 + (v_B)_2}{4} = 0.6$

$\dfrac{1}{2}(2)(1.75)^2 - 7.848 x = 0$

15-15. $\quad 0 = [v_C \sin 20°]^2 + 2(-9.81)(4 - 1)$

$3.5(10^3)(21.08) + 0 = 7(10^3)v_2$

$4 = 0 + 0 + \frac{1}{2}(9.81)t^2$

$(10.54)(0.903)$

15-16. $4(4) + 0 = 4(v_A)_2 + 4(v_B)_2$

$0.7 = \dfrac{(v_B)_2 - (v_A)_2}{4}$

$\dfrac{1}{2}(4)(3.40)^2 + 0 = 0 + \dfrac{1}{2}(500)x^2$

15-17. $\dfrac{13.2(10^{-3})}{32.2}(2\sin 30°) - \dfrac{6.6(10^{-3})}{32.2}(3\cos 60°) = \dfrac{13.2(10^{-3})}{32.2}(v_A)_{2x} + \dfrac{6.6(10^{-3})}{32.2}(v_B)_{2x}$

$0.65 = \dfrac{(v_B)_{2x} - (v_A)_{2x}}{2\sin 30° + 3\cos 60°}$

$\dfrac{13.2(10^{-3})}{32.2} 2\cos 30° = \dfrac{13.2(10^{-3})}{32.2}(v_A)_{2y}$

$\dfrac{-6.6(10^{-3})}{32.2} 3\sin 60° = \dfrac{-6.6(10^{-3})}{32.2}(v_B)_{2y}$

15-18. $0 = (353.5)^2 + 2(-9.81)(h - 0)$

$(6369.1)(3)(353.5)$

15-19. $0 + (3)(20)(0.4) - 10(3)(0.4) = \left(\dfrac{20}{32.2}\right)(v_b - 2)(0.4) + \left(\dfrac{10}{32.2}\right)(v_b)(0.4)$

15-20. $2[(0.5)(2)(0.3)] + 0.8(4) = 2[(0.5)(v_2)(0.3)]$

15-21. $150(19.44)(60) = 150(v_B)\cos\theta(57)$

$\frac{1}{2}(150)(19.44)^2 + 150(9.81)h = \frac{1}{2}(150)(v_B)^2$

15-22. $\frac{1}{2}\left(\frac{80}{32.2}\right)(8)^2 + 0 = \frac{1}{2}\left(\frac{80}{32.2}\right)(v_B)^2 - 80(2)$

$(10)\left(\frac{80}{32.2}\right)(8) = 7\left(\frac{80}{32.2}\right)(v_B)_{\text{horiz.}}$

16-1. $50(0.5)$

$25/0.2$

$125 = 0 + 2t$

16-2. $\int_5^\omega \omega\, d\omega = \int_0^{4\pi} 0.2\theta\, d\theta$

$(2.5)(7.52)$

16-3. $1.2(250)$

$(1.2)^2(200)$

16-4. $6(50\sqrt{2})/50$

$8.49(0.05\sqrt{2})$

16-5. $\int_0^\omega \omega\, d\omega = \int_0^\theta 4\theta\, d\theta$

$12(8\cos 30°)$

16-6. $1.5\cos\theta$

$-1.5\sin\theta\,\dot\theta$

$-1.5\cos\theta(\dot\theta)^2 - 1.5\sin\theta\,\ddot\theta$

16-7. $2\cot\theta$

$-2\csc^2\theta\,\dot\theta$

16-8. $2r\theta$
$2r\omega$

16-9. $4\sin\theta$
$4\cos\theta\,\dot\theta$
$4\cos\theta$
$-4\sin\theta\,\dot\theta$

16-10. $2\cos\theta$
$-2\sin\theta\,\omega$
$2\sin\theta$
$2\cos\theta\,\omega$

16-11. $200\mathbf{j} + (\omega\mathbf{k}) \times (25\mathbf{j})$

16-12. $-\omega_{AB}(2)\mathbf{j} = -4\mathbf{j} - 3\omega_{BC}\sin30°\mathbf{i} + 3\omega_{BC}\cos30°\mathbf{j}$

16-13. $-\left(\dfrac{4}{5}\right)v_B\mathbf{i} - \left(\dfrac{3}{5}\right)v_B\mathbf{j} = -8\mathbf{i} + (\omega_{BC}\mathbf{k}) \times (-5\mathbf{i})$

16-14. $v_A\mathbf{i} = -6\mathbf{j} + (\omega\mathbf{k}) \times (3\mathbf{i} - 4\mathbf{j})$

16-15. $-v_C\mathbf{i} = -600\mathbf{i} + (\omega_{BC}\mathbf{k}) \times (100\cos30°\mathbf{i} + 100\sin30°\mathbf{j})$

16-16. $60/1.25$
$48(25)$

16-17. $7(2)$
$(5.39)(2)$
$2/5$

16-18. $5(2)$

$(10)/(4 + 2)$

16-19. $\omega(r_2 - r_1)$

ωr_2

16-20. $10/\tan 9.46°$

$18/60$

$(10/\sin 9.46°)(0.3)$

16-21. $4/0.5\sqrt{2}$

$5.657(0.5)$

$\alpha_{AB}(0.5)\mathbf{i} - (5.657)^2(0.5)\mathbf{j} = 3\cos 45°\mathbf{i} - 3\sin 45°\mathbf{j} + (\alpha_{BC}\mathbf{k}) \times (-0.5\mathbf{i}) - (5.657)^2(-0.5\mathbf{i})$

16-22. $(a_C)_x\mathbf{i} - 4\mathbf{j} = -(a_D)_y\mathbf{j} + (\alpha\mathbf{k}) \times (-0.075\mathbf{i}) - (40)^2(-0.075\mathbf{i})$

16-23. $(a_A)_x\mathbf{i} - (a_A)_y\mathbf{j} = 0.6\mathbf{i} + (-4\mathbf{k}) \times (0.15\mathbf{j}) - (2)^2(0.15\mathbf{j})$

$(a_B)\mathbf{i} = 1.20\mathbf{i} - 0.6\mathbf{j} + (\alpha_{AB}\mathbf{k}) \times (0.4\mathbf{i} - 0.3\mathbf{j}) - \mathbf{0}$

16-24. $-1.20\mathbf{j} = a_A\mathbf{i} + (\alpha_{AB}\mathbf{k}) \times (-1.47\mathbf{i} + 0.3\mathbf{j}) - \mathbf{0}$

16-25 $-(a_C)_x\mathbf{i} - 0.122\mathbf{j} = -1.80\cos 30°\mathbf{i} + 1.80\sin 30°\mathbf{j} - 0.225\sin 30°\mathbf{i} - 0.225\cos 30°\mathbf{j}$

$+ (\alpha_{BC}\mathbf{k}) \times (0.2\mathbf{i}) - (1.125)^2(0.2\mathbf{i})$

17-1. $0.2N_B = 1500 a_G$

$N_A + N_B - 1500(9.81) = 0$

$N_A(1.25) - N_B(0.75) + (0.2N_B)(0.35) = 0$

17-2. $A_y - 10 = -0.3106(4)\sin 30°$
$V_x = 0.3106(4)\cos 30°$
$M_A = 0.3106(4)\cos 30°(1)$

17-3. $1.8v^2 = 1300 a_G$
$(0.7\overline{5})(1.8v^2) - 1.25 N_A = 0$
$N_A - (1300)(9.81) = 0$

17-4. $N_b - 98.1 = -10(6)$
$0.4(38.1) = 10\alpha(1.5)$

17-5. $0.4 N_C = \dfrac{150}{32.2} a \cos 30°$

$N_C - 150 = \dfrac{150}{32.2} a \sin 30°$

17-6. $A_x = 4000(3) + 800(3)$
$14\,400(1.75) + A_y(8) - 4000(9.81)(4) - 800(9.81)(2.5) = 4000(3)(1.5) + 800(3)(1.75)$

17-7. $15(1.5) = \left[\dfrac{1}{3}\left(\dfrac{15}{32.2}\right)(3)^2\right]\alpha$

$A_x = \dfrac{15}{32.2}(5)^2(1.5)$

$A_y - 15 = \dfrac{-15}{32.2}(16.1)(1.5)$

17-8. $30(1.25) = \left[\dfrac{400}{32.2}(1.30)^2\right]\alpha$

17-9. $A_x = \dfrac{30}{32.2}\left(\sqrt{2}\,\alpha\right)\left(\dfrac{1}{\sqrt{2}}\right)$

$A_y - 30 = -\dfrac{30}{32.2}\left(\sqrt{2}\,\alpha\right)\left(\dfrac{1}{\sqrt{2}}\right)$

$$30(1) = \frac{30}{32.2}(\sqrt{2}\alpha)\sqrt{2} + \left[\frac{1}{12}\left(\frac{30}{32.2}\right)[(2)^2 + (2)^2]\right]\alpha$$

17-10. $\quad 0.3N_A(0.15) = \left[\frac{1}{2}(20)(0.15)^2\right]\alpha$

$F_{CB}\sin 30° - N_A = 0$

$F_{CB}\cos 30° - 196.2 + 0.3N_A = 0$

17-11. $\quad \mu_A N_A = 80(2.5\alpha)\cos\theta - 80(2.5\omega^2)\sin\theta$

$N_A - 784.8 = -80(2.5\omega^2)\cos\theta - 80(2.5\alpha)\sin\theta$

$784.8(2.5\sin\theta) = [80(1.2)^2]\alpha + 80(2.5\alpha)(2.5)$

17-12. $\quad 30\sin\theta - F_W = (0.9317)a_G$

$N_W - 30\cos\theta = 0$

$F_W(0.5) = [(0.9317)(0.23)^2]\alpha$

17-13. $\quad 30 = 200 a_G$

$T - 1962 = 0$

$30(2) = \left[\frac{1}{12}(200)(4)^2\right]\alpha$

17-14. $\quad F_A = \frac{180}{32.2}(3.04\alpha)\left(\frac{3}{3.04}\right)$

$N_A - 180 = -\frac{180}{32.2}(3.04\alpha)\left(\frac{0.5}{3.04}\right)$

$180(0.5) = \left[\frac{180}{32.2}(2.2)^2\right]\alpha + \frac{180}{32.2}(3.04\alpha)(3.04)$

17-15. $\quad 30 + F_A = \frac{80}{32.2}a_G$

$N_A - 80 = 0$

$-F_A(2) + 30(0.5) = \left[\frac{80}{32.2}(0.75)^2\right]\alpha$

170 Answers

17-16. $1500 - T_{BC} - 0.3N_A = 100(a_G)_x$

$1500(2\sin 45°) - 0.3N_A(2\sin 45°) - N_A(2\cos 45°) + T_{BC}(2\sin 45°) = \left[\frac{1}{12}(100)(4)^2\right]\alpha$

$N_A - 981 = 100(a_G)_y$

18-1. $0 + 800(25) = \frac{1}{2}\left[\frac{1}{3}\left(\frac{800}{32.2}\right)(50)^2\right]\omega^2$

18-2. $(0+0) + 120\left(\frac{4}{1.5}\right) - 15(4) = \left[\frac{1}{2}\left(\frac{15}{32.2}\right)(v)^2\right] + \left[\frac{1}{2}\left(\frac{30}{32.2}\right)(0.8)^2\right]\left(\frac{v}{1.5}\right)^2$

18-3. $0 + (400)(8) = \frac{1}{2}[200(0.325)^2](\omega_2)^2$

18-4. $0 + 80(10^3)(\pi) - 180(120) = \frac{1}{2}\left[\frac{15\,000}{32.2}(37)^2\right]\omega^2 + \frac{1}{2}\left(\frac{180}{32.2}\right)(60\omega)^2$

18-5. $0 + (150)(2.5 - 1.75) = \frac{1}{2}\left(\frac{150}{32.2}\right)v_G^2$

18-6. $0 + 3\left(\left(\frac{10}{12}\right) - \left(\frac{10}{12}\right)\cos 30°\right) = \frac{1}{2}\left(\frac{3}{32.2}\right)(v_G)^2 + \frac{1}{2}\left[\frac{2}{5}\left(\frac{3}{32.2}\right)\left(\frac{1.5}{12}\right)^2\right]\omega^2$

18-7. $0 + (0.5)(9.81)(0.05) = \frac{1}{2}\left[\frac{1}{12}(0.5)(0.2)^2\right]\omega_{AB}^2 + \frac{1}{2}(0.5)(v_G)^2 + (0.5)(9.81)(0.0268)$

18-8. $[0+0] + 400(10\sin 45°) + 1800(15\sin 45°)$
$- 2400(12.5\sin 45°) =$

$\frac{1}{2}\left[\frac{1}{12}\left(\frac{400}{32.2}\right)(40)^2 + \frac{400}{32.2}(10)^2\right]\omega^2 +$

$\frac{1}{2}\left[\frac{1}{12}\left(\frac{1800}{32.2}\right)(30)^2 + \frac{1800}{32.2}(15)^2\right]\omega^2 +$

$\frac{1}{2}\left[\frac{1}{12}\left(\frac{2400}{32.2}\right)(5)^2 + \left(\frac{2400}{32.2}\right)(12.5)^2\right]\omega^2$

18-9. $(0 + 0) + 0 = \frac{1}{2}(4)(v_A)^2 + \frac{1}{2}[2(0.05)^2]\omega^2 - 4(9.81)(1)$

18-10. $0 + 0 = \frac{1}{2}\left[\frac{1}{12}(5)(0.8)^2\right]\omega^2 + \frac{1}{2}(5)(0.2\omega)^2 - 49.05(0.2)$

$O_x = 5(0.2\alpha)$
$O_y - 49.05 = 5(6.48)^2(0.2)$
$0 = I_o \alpha;$

19-1. $0 + 2(0.6)(4) + F_A(0.6)(4) = \left[\frac{1}{2}\left(\frac{50}{32.2}\right)(0.6)^2\right]\omega$

$0 + 2(4) - F_A(4) = \frac{50}{32.2} v_G$

19-2. $0 + 2(9.81)(3)(0.08) = \left(\frac{1}{2}(2)(0.08)^2\right)\omega + 2(v_G)(0.08)$

19-3. $0 - 3(F)(3)(0.04) + (0.4)(3) = [0.8(0.031)^2]\omega_A$
$0 + F(3)(0.02) = [0.3(0.015)^2]\omega_B$

19-4. $[(0.0466)(0.13)^2](20) + M(2.5) - F(2.5)(0.2) =$
$\qquad [(0.0466)(0.13)^2](40)$

$(0.0248)(4) + F(2.5) - (0.8)(0.3)(2.5) = 0.0248(8)$

19-5. $P(1) + 0.4N_W(0.4) - N_W(0.5) = 0$

19-6. $0 + \int_0^3 12t\, dt + 0.125\int T_1 dt - 0.125 \int T_2 dt =$
$\qquad [(30)(0.095)^2](\omega_A)_2$

$0 + (0.125)\int T_2 dt - (0.125)\int T_1 dt =$
$\qquad \left[\frac{1}{2}(25)(0.125)^2\right](\omega_D)_2$

172 Answers

19-7. $0 + 0.2\,N_C(t) - T_{AB}(t) = 0$

$0 + N_C(t) - 490.5(t) = 0$

$\left[\frac{1}{2}(50)(0.2)^2\right]30 - 0.2\,N_C(t)(0.2) = 0$

19-8. $0 + 0 = -\left(\frac{300}{32.2}(8)^2\right)\omega + 10\left[\left(\frac{150}{32.2}\right)(4 - 10\omega)\right]$

19-9. $\frac{1}{6}ma^2(\omega_1) = \frac{2}{3}ma^2\omega_2$

19-10. $0 + (15)(9.81)(1.5) =$

$\frac{1}{2}\left[\frac{1}{3}(15)(3)^2\right]\omega_2^2 + (15)(9.81)(1.5)\sin 60°$

$\left[\frac{1}{12}(15)(3)^2\right](1.146) + 15(1.720)\left(\frac{3}{2} - \frac{0.5}{\sin 60°}\right)$

$= \left[\frac{1}{12}(15)(3)^2\right]\omega_3 + 15(v_G)_3\left[\frac{3}{2} - \frac{0.5}{\sin 60°}\right]$

19-11. $0 + 3(1.5) = \frac{1}{2}\left(\frac{3}{32.2}\right)(v_G)^2$

$\left(\frac{3}{32.2}\right)(9.83)(1) = \left[\frac{1}{3}\left(\frac{3}{32.2}\right)(2)^2\right]\omega_2$

$\frac{1}{2}\left[\frac{1}{3}\left(\frac{3}{32.2}\right)(2)^2\right](7.37)^2 + 0 = \frac{1}{2}\left[\frac{1}{3}\left(\frac{3}{32.2}\right)(2)^2\right](\omega_3)^2 - 3(1)$